Geometry

Applying • Reasoning • Measuring

Chapter 9
Resource Book

The Resource Book contains the wide variety of blackline masters available for Chapter 9. The blacklines are organized by lesson. Included are support materials for the teacher as well as practice, activities, applications, and assessment resources.

McDougal Littell
A HOUGHTON MIFFLIN COMPANY
Evanston, Illinois • Boston • Dallas

Contributing Authors

The authors wish to thank the following individuals for their contributions to the Chapter 9 Resource Book.

Eric J. Amendola
Karen Collins
Michael Downey
Patrick M. Kelly
Edward H. Kuhar
Lynn Lafferty
Dr. Frank Marzano
Wayne Nirode
Dr. Charles Redmond
Paul Ruland

ISBN: 0-618-02072-1

123456789-VEI- 04 03 02 01 00

Contents

9   *Right Triangles and Trigonometry*

Contents

Contents

Descriptions of Resources

This Chapter Resource Book is organized by lessons within the chapter in order to make your planning easier. The following materials are provided:

Tips for New Teachers These teaching notes provide both new and experienced teachers with useful teaching tips for each lesson, including tips about common errors and inclusion.

Parent Guide for Student Success This guide helps parents contribute to student success by providing an overview of the chapter along with questions and activities for parents and students to work on together.

Prerequisite Skills Review Worked-out examples are provided to review the prerequisite skills highlighted on the Study Guide page at the beginning of the chapter. Additional practice is included with each worked-out example.

Strategies for Reading Mathematics The first page teaches reading strategies to be applied to the current chapter and to later chapters. The second page is a visual glossary of key vocabulary.

Lesson Plans and Lesson Plans for Block Scheduling This planning template helps teachers select the materials they will use to teach each lesson from among the variety of materials available for the lesson. The block-scheduling version provides additional information about pacing.

Warm-Up Exercises and Daily Homework Quiz The warm-ups cover prerequisite skills that help prepare students for a given lesson. The quiz assesses students on the content of the previous lesson. (Transparencies also available)

Activity Support Masters These blackline masters make it easier for students to record their work on selected activities in the Student Edition.

Alternative Lesson Openers An engaging alternative for starting each lesson is provided from among these four types: *Application, Activity, Geometry Software,* or *Visual Approach.* (Color transparencies also available)

Technology Activities with Keystrokes Keystrokes for Geometry software and calculators are provided for each Technology Activity in the Student Edition, along with alternative Technology Activities to begin selected lessons.

Practice A, B, and C These exercises offer additional practice for the material in each lesson, including application problems. There are three levels of practice for each lesson: A (basic), B (average), and C (advanced).

Contents

Reteaching with Additional Practice These two pages provide additional instruction, worked-out examples, and practice exercises covering the key concepts and vocabulary in each lesson.

Quick Catch-Up for Absent Students This handy form makes it easy for teachers to let students who have been absent know what to do for homework and which activities or examples were covered in class.

Cooperative Learning Activities These enrichment activities apply the math taught in the lesson in an interesting way that lends itself to group work.

Interdisciplinary Applications/Real-Life Applications Students apply the mathematics covered in each lesson to solve an interesting interdisciplinary or real-life problem.

Math and History Applications This worksheet expands upon the Math and History feature in the Student Edition.

Challenge: Skills and Applications Teachers can use these exercises to enrich or extend each lesson.

Quizzes The quizzes can be used to assess student progress on two or three lessons.

Chapter Review Games and Activities This worksheet offers fun practice at the end of the chapter and provides an alternative way to review the chapter content in preparation for the Chapter Test.

Chapter Tests A, B, and C These are tests that cover the most important skills taught in the chapter. There are three levels of test: A (basic), B (average), and C (advanced).

SAT/ACT Chapter Test This test also covers the most important skills taught in the chapter, but questions are in multiple-choice and quantitative-comparison format. (See *Alternative Assessment* for multi-step problems.)

Alternative Assessment with Rubrics and Math Journal A journal exercise has students write about the mathematics in the chapter. A multi-step problem has students apply a variety of skills from the chapter and explain their reasoning. Solutions and a 4-point rubric are included.

Project with Rubric The project allows students to delve more deeply into a problem that applies the mathematics of the chapter. Teacher's notes and a 4-point rubric are included.

Cumulative Review These practice pages help students maintain skills from the current chapter and preceding chapters.

LESSON 9.1

TEACHING TIP Have students use the physical models for similar triangles from the Activity on page 527 when looking at Theorem 9.1 and solving problems similar to Example 1 on page 528. Alternatively, you could encourage them to draw individual diagrams, like those shown in the solution of Example 1, part a.

TEACHING TIP Consider having extra index cards available for students to make additional models like those in the Activity. They can keep one set of triangles unmarked. A second set of triangles could be labeled and used to explore the relationships of the Geometric Mean Theorems on page 529.

COMMON ERROR Students usually have more trouble writing the proportion for problems like part b in Example 2 on page 530 than for problems like part a. Have a variety of examples of each type to provide the practice of just writing the proportion. They could also refer to their index card models.

LESSON 9.2

TEACHING TIP Students are usually familiar with the Pythagorean Theorem or formula by the time they are studying geometry. Some may not be aware of why the variables a, b, and c are typically used in the statement of the theorem. Draw a right triangle like the one shown in Theorem 9.4 on page 535, label the vertices with A, B, and C to correspond to the labels of the sides, and tell students that this is a common way to label the vertices of a right triangle.

TEACHING TIP Notice the difference between the diagram used in Theorem 9.4 and the one used below the theorem, in the plan for its proof. The triangles have the same orientation, but different labeling. This may be confusing to students. Be ready to answer questions that may arise due to the differences in labeling. Remind students that any labels could be used as long as the accuracy of the concept or statement of the theorem is preserved.

INCLUSION After discussing Example 1 on page 536, consider showing additional multiples of the 3-4-5 and 5-12-13 right triangles. Have students guess what the next multiple is and have them check to see if the numbers form a Pythagorean triple. You can also give the measures of two legs and have students guess the hypotenuse of the multiple. Students with limited English proficiency can benefit from this pattern recognition process.

TEACHING TIP Prior to Example 3 on page 537, review with students that the altitude drawn from the vertex angle to the base in an isosceles triangle has special properties. It is the perpendicular bisector of the base, which is the concept needed for this problem.

LESSON 9.3

TEACHING TIP Activity 9.3, on page 542, leads to the statements of Theorems 9.6 and 9.7 on page 544. Stress that students should always identify the longest side of a given triangle first before applying the triangle inequalities in these theorems or even when using the Pythagorean Theorem.

LESSON 9.4

TEACHING TIP The theorems about special right triangles on page 551 involve the use of radicals. Answers to problems may be given in simplest radical form (as in Examples 1, 2, and 3) or a decimal approximation (as in Example 5 and 6). Discuss the possibilities with students and how you would like them to round for the approximations when they solve related problems.

LESSON 9.5

TEACHING TIP Students will work with trigonometric ratios in Lessons 9.5 and 9.6. They should realize that trigonometry enables them to determine all the measures of angles and sides in a right triangle when only two pieces of information about the triangles are given.

Chapter Support

TEACHING TIP Point out to students that the angle symbol is not used when writing ratios or equations involving trigonometric ratios (functions). While looking at Example 1 on page 558, explain that the value of a trigonometric ratio is frequently expressed as a decimal approximation, rounded to the nearest ten thousandth (four decimal places). Compare this fact to the trigonometry tables on page 845 and to the result that a student would get when using a calculator.

COMMON ERROR The hypotenuse of a right triangle is usually the easiest side to identify in a diagram. Students tend to make mistakes when trying to identify the other two sides by confusing which is opposite and which is adjacent. Stress that students identify the hypotenuse first. Then they should focus on a particular angle in the right triangle and identify the side opposite that angle. The remaining side must be the adjacent one, if everything else was identified correctly. This is illustrated in the solution to Example 2 on page 559.

LESSON 9.6

TEACHING TIP Students need to practice interpreting and using the inverse trigonometric functions on their calculators. Give them decimal values and a function and ask them to find the angle measure. You should also have them practice using just the trig tables. You may want to do this following the Activity on page 567 and prior to Example 1 on page 568.

LESSON 9.7

COMMON ERROR As in Lesson 7.4, students need to be reminded about the proper use of vector notation: the special brackets and the vector arrow.

TEACHING TIP Use Example 2 on page 574 to make students aware of how to write their answers regarding magnitude and direction for problems involving vectors. The written answer to part b should serve as a good model.

TEACHING TIP The diagrams of vector addition on page 575 may be confusing to students. Carefully explain and draw the diagram in Example 4 in stages, using the given information. Students need to understand the difference between that diagram and the one presented in Example 5.

Outside Resources

BOOKS/PERIODICALS
Vonder Embse, Charles and Arne Engebretsen. "Using Interactive-Geometry Software for Right-Angle Trigonometry." *Mathematics Teacher* (October 1996); pp 602–605.

ACTIVITIES/MANIPULATIVES
Waiveris, Charles and Timothy V. Craine. "Activities: Where Are We?" *Mathematics Teacher* (September 1996); pp. 524–534.

SOFTWARE
Exploring Trigonometry with The Geometer's Sketchpad. Blackline masters and Macintosh and Windows disks with sample sketches and scripts. Berkeley, CA. Key Curriculum Press.

VIDEOS
Apostol, Tom M. *The Theorem of Pythagoras.* Includes program guide/workbook that includes exercises. Reston, VA; NCTM.

NAME _____ DATE _____

Parent Guide for Student Success

For use with Chapter 9

Chapter Overview One way that you can help your student succeed in Chapter 9 is by discussing the lesson goals in the chart below. When a lesson is completed, ask your student to interpret the lesson goals for you and to explain how the mathematics of the lesson relates to one of the key applications listed in the chart.

Lesson Title	Lesson Goals	Key Applications
9.1: Similar Right Triangles	Solve problems involving similar right triangles formed by the altitude drawn to the hypotenuse. Use a geometric mean to solve problems.	• Roof Height • Monorail Station • Rock Climbing
9.2: The Pythagorean Theorem	Prove the Pythagorean Theorem. Use the Pythagorean Theorem to solve real-life problems.	• Malaysian Support Beam • Softball Diamond • Trans-Alaska Pipeline
9.3: The Converse of the Pythagorean Theorem	Use the Converse of the Pythagorean Theorem to solve problems. Use side lengths to classify triangles by their angle measures.	• House Foundation • Babylonian Tablet • Air Travel
9.4: Special Right Triangles	Find the side lengths of special right triangles. Use special right triangles to solve real-life problems.	• Tipping Platform • Jewelry • Tools
9.5: Trigonometric Ratios	Find the sine, the cosine, and the tangent of an acute angle. Use trigonometric ratios to solve real-life problems.	• Forestry • Water Slide • Lunar Cartography
9.6: Solving Right Triangles	Solve a right triangle. Use right triangles to solve real-life problems.	• Space Shuttle • Longs Peak • Wheelchair Ramps
9.7: Vectors	Find the magnitude and the direction of a vector. Add vectors.	• Velocity of a Jet • Tug-of-War Game • Skydiving

Study Strategy

List What You Know is the study strategy featured in Chapter 9 (see page 526). Have your student list what he or she already knows about the title of the chapter and what he or she expects to learn. Encourage your student to review the list when the chapter is finished. Discuss what your student learned and whether or not his or her expectations were met.

NAME _____ DATE _____

Parent Guide for Student Success

For use with Chapter 9

Key Ideas Your student can demonstrate understanding of key concepts by working through the following exercises with you.

Lesson	Exercise
9.1	To find the height of a tree, you hold a cardboard square at your eye level and line up consecutive edges of the square with the top and bottom of the tree. Your eye is 5 feet above the ground and 14 feet from the tree (perpendicular distance). Estimate the height of the tree.
9.2	A right triangle has a hypotenuse that is 12 centimeters long and a leg that is 10 centimeters long. Find the approximate area of the triangle.
9.3	The ancient Egyptians used a closed loop of rope, knotted to form evenly spaced sections, and the idea behind the Converse of the Pythagorean Theorem to determine right angles when building the pyramids. If a loop of rope had 24 sections, how long should each side be to form a right triangle?
9.4	When a 30-foot drawbridge is raised 30°, its outer end is just over the outer edge of the castle's moat. What is the horizontal distance from the inner end of the drawbridge to the outer edge of the moat?
9.5	A road rises 60 feet in 1000 feet of horizontal distance. Find the tangent, sine, and cosine of the angle of elevation to 4 decimal places. (*Hint:* Use the Pythagorean Theorem.)
9.6	In $\triangle QRS$, $m\angle R = 90°$, $m\angle S = 16°$, and $QR = 7$ in. Solve the right triangle. Round decimals to the nearest tenth.
9.7	Vector PQ has terminal points $P(2, -1)$ and $Q(-3, 2)$. Write the component form of the vector and find its magnitude.

Home Involvement Activity

You will need: A piece of rope about 7 feet long

Directions: Divide the rope into 12 equal sections by making 11 knots to mark off the sections. Use the knotted rope to construct a right angle as follows: One person holds the two ends of the rope together. The second person holds the rope at the fifth knot. A third person pulls the rope taut at the eighth knot. The angle formed at the eighth knot should be a right angle. How do you know that this angle is a right angle?

Answers

9.1: about 44 ft 9.2: about 33.2 cm² 9.3: 6, 8, and 10 spaces 9.4: $15\sqrt{3}$ or almost 26 ft 9.5: 0.0600, 0.0599, 0.9982 9.6: $RS \approx 24.4$ in., $QS \approx 25.4$ in., $m\angle Q = 74°$ 9.7: $(-5, 3)$, about 5.8

Geometry
Chapter 9 Resource Book

NAME _____ DATE _____

Prerequisite Skills Review

For use before Chapter 9

EXAMPLE 1

Simplifying Radical Expressions

Write answers in simplest radical form.

a. $\sqrt{48}$ **b.** $\sqrt{20}$

SOLUTION

a. $\sqrt{48} = \sqrt{16} \cdot \sqrt{3}$ **b.** $\sqrt{20} = \sqrt{4} \cdot \sqrt{5}$

$\quad\quad\quad = 4 \cdot \sqrt{3}$ $= 2 \cdot \sqrt{5}$

$\quad\quad\quad = 4\sqrt{3}$ $= 2\sqrt{5}$

Exercises for Example 1

Simplify each radical expression. Write answers in simplest radical form.

1. $\sqrt{72}$ **2.** $\sqrt{80}$ **3.** $\sqrt{45}$

4. $\sqrt{32}$ **5.** $\sqrt{90}$ **6.** $\sqrt{52}$

EXAMPLE 2

Solving Proportions

Solve each proportion.

a. $\dfrac{x}{6} = \dfrac{x+2}{4}$ **b.** $\dfrac{y-3}{3} = \dfrac{y}{4}$

SOLUTION

a. $\dfrac{x}{6} = \dfrac{x+2}{4}$ **b.** $\dfrac{y-3}{3} = \dfrac{y}{4}$

       $6x + 12 = 4x$ $3y = 4y - 12$

           $12 = -2x$ $-y = 12$

           $-6 = x$ $y = -12$

Exercises for Example 2

Solve each proportion.

7. $\dfrac{x}{5} = \dfrac{x-4}{3}$ **8.** $\dfrac{x}{12} = \dfrac{x+3}{6}$ **9.** $\dfrac{x+3}{7} = \dfrac{x-1}{3}$

10. $\dfrac{x-6}{2} = \dfrac{x}{6}$ **11.** $\dfrac{x+5}{8} = \dfrac{x}{4}$ **12.** $\dfrac{1}{3} = \dfrac{2x+1}{x+8}$

Prerequisite Skills Review

For use before Chapter 9

EXAMPLE 3 ## Component Form of Vectors

Write the component form of $\overrightarrow{AB}$.

a. $A(3, 2), B(5, 6)$ **b.** $A(-7, 1), B(2, -5)$

SOLUTION

a. $\overrightarrow{AB} = \langle 5 - 3, 6 - 2 \rangle$ **b.** $\overrightarrow{AB} = \langle 2 - (-7), -5 - 1 \rangle$
   $= \langle 2, 4 \rangle$ $= \langle 9, -6 \rangle$

Exercises for Example 3

Write the component form of $\overrightarrow{AB}$.

13. $A(0, 0), B(3, 6)$ **14.** $A(7, 1), B(0, 6)$ **15.** $A(4, 4), B(6, 6)$

16. $A(3, -5), B(-2, -6)$ **17.** $A(9, 2), B(-1, -1)$ **18.** $A(-2, 4), B(1, 8)$

Strategies for Reading Mathematics

For use with Chapter 9

Strategy: Reading Tables

You will find two tables at the back of your book that you can use as you study right triangles and trigonometry. The Table of Squares and Square Roots can be found on page 844 and the Table of Trigonometric Ratios on page 845.

Table of Squares and Square Roots

No.	Square	Sq. Root
1	1	1.000
2	(4)	1.414
3	9	1.732
4	16	2.000

←The square root of 2 is ④.
←The square root of 3 is 1.732 .

Each row shows a number, its square, and its square root.

Table of Trigonometric Ratios

Angle	Sine	Cosine	Tangent
1°	.0175	.9998	.0715
2°	.0349	.9994	.0349
3°	.0523	.9986	.0524

←The sine of 1° is .0175 .

Each row shows an angle measure, its sine, cosine, and tangent.

STUDY TIP

Using Tables

Sometimes it is difficult to keep your place when you are looking at rows in the middle of a table. You can keep your place in the row by putting a straightedge or the edge of a piece of paper under the row.

Questions

1. Use the Table of Squares and Square Roots above to find the square root of 2.

2. Use the Table of Trigonometric Ratios above to find the sine of 2°.

3. Use the Table of Trigonometric Ratios above to find the tangent of 3°.

4. Explain how to use the Table of Trigonometric Ratios above to find the measure of the angle whose tangent is 0.0349. Then find the measure of the angle.

NAME _____ DATE _____

Strategies for Reading Mathematics

For use with Chapter 9

Visual Glossary

The Study Guide on page 526 lists the key vocabulary for Chapter 9 as well as review vocabulary from previous chapters. Use the page references on page 526 or the Glossary in the textbook to review key terms from prior chapters. Use the visual glossary below to help you understand some of the key vocabulary in Chapter 9. You may want to copy these diagrams into your notebook and refer to them as you complete the chapter.

GLOSSARY

Pythagorean triple (p. 536) A set of three positive integers a, b, and c that satisfy the equation $c^2 = a^2 + b^2$.

trigonometric ratio (p. 558) A ratio of the lengths of two sides of a right triangle.

sine (p. 558) A trigonometric ratio, abbreviated as *sin*. For right triangle *ABC*, the sine of acute angle *A* is

$$\sin A = \frac{\text{side opposite } \angle A}{\text{hypotenuse}}.$$

cosine (p. 558) A trigonometric ratio, abbreviated as *cos*. For right triangle *ABC*, the cosine of acute angle *A* is

$$\cos A = \frac{\text{side adjacent to } \angle A}{\text{hypotenuse}}.$$

tangent (p. 558) A trigonometric ratio, abbreviated as *tan*. For right triangle *ABC*, the tangent of acute angle *A* is

$$\tan A = \frac{\text{side opposite } \angle A}{\text{side adjacent to } \angle A}.$$

Pythagorean Triples and the Pythagorean Theorem

The Pythagorean Theorem is used to find and verify Pythagorean triples. The Pythagorean Theorem states that in a right triangle, the square of the length of the hypotenuse is equal to the sum of the squares of the lengths of the legs.

$$c^2 = a^2 + b^2$$
$$\downarrow \quad \downarrow \quad \downarrow$$
$$10^2 = 8^2 + 6^2$$

$$100 = 64 + 36$$

So 10, 8, and 6 form a Pythagorean triple.

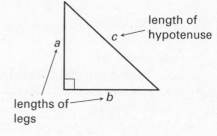

length of hypotenuse

lengths of legs

Finding Trigonometric Ratios

The three basic trigonometric ratios are sine, cosine, and tangent. You can find the sine, cosine, or tangent of either acute angle of any right triangle.

$$\sin A = \frac{a}{c} \qquad \cos A = \frac{b}{c} \qquad \tan A = \frac{a}{b}$$

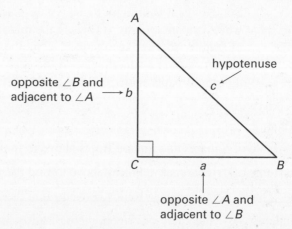

opposite $\angle B$ and adjacent to $\angle A$ → b

hypotenuse

opposite $\angle A$ and adjacent to $\angle B$

$$\sin B = \frac{b}{c} \qquad \cos B = \frac{a}{c} \qquad \tan B = \frac{b}{a}$$

Geometry
Chapter 9 Resource Book

TEACHER'S NAME _____ CLASS _____ ROOM _____ DATE _____

Lesson Plan

1-day lesson (See *Pacing the Chapter*, TE pages 524C–524D)   **For use with pages 527–534**

GOALS 1. **Solve problems involving similar right triangles formed by the altitude drawn to the hypotenuse of a right triangle.**
2. **Use a geometric mean to solve problems.**

State/Local Objectives _____

_____

✓ **Check the items you wish to use for this lesson.**

STARTING OPTIONS
_____ Prerequisite Skills Review: CRB pages 5–6
_____ Strategies for Reading Mathematics: CRB pages 7–8
_____ Homework Check: TE page 509: Answer Transparencies
_____ Warm-Up or Daily Homework Quiz: TE pages 527 and 513, CRB page 11, or Transparencies

TEACHING OPTIONS
_____ Motivating the Lesson: TE page 528
_____ Lesson Opener (Geometry Software): CRB page 12 or Transparencies
_____ Technology Activity with Keystrokes: CRB page 13
_____ Examples 1–3: SE pages 528–530
_____ Extra Examples: TE pages 528–530 or Transparencies
_____ Closure Question: TE page 530
_____ Guided Practice Exercises: SE page 531

APPLY/HOMEWORK
Homework Assignment
_____ Basic   11–30, 33, 34, 41, 42, 44–50 even
_____ Average   11–30, 33–36, 41, 42, 44–50 even
_____ Advanced   11–30, 32–36, 41–43, 44–50 even

Reteaching the Lesson
_____ Practice Masters: CRB pages 14–16 (Level A, Level B, Level C)
_____ Reteaching with Practice: CRB pages 17–18 or Practice Workbook with Examples
_____ Personal Student Tutor

Extending the Lesson
_____ Applications (Interdisciplinary): CRB page 20
_____ Challenge: SE page 534; CRB page 21 or Internet

ASSESSMENT OPTIONS
_____ Checkpoint Exercises: TE pages 528–530 or Transparencies
_____ Daily Homework Quiz (9.1): TE page 534, CRB page 24, or Transparencies
_____ Standardized Test Practice: SE page 534; TE page 534; STP Workbook; Transparencies

Notes _____

_____

_____

Geometry
Chapter 9 Resource Book

9

TEACHER'S NAME _____ CLASS _____ ROOM _____ DATE _____

Lesson Plan for Block Scheduling

Half-day lesson (See *Pacing the Chapter,* TE pages 524C–524D)   For use with pages 527–534

GOALS 1. **Solve problems involving similar right triangles formed by the altitude drawn to the hypotenuse of a right triangle.**
2. **Use a geometric mean to solve problems.**

State/Local Objectives _____

_____

_____

✓ **Check the items you wish to use for this lesson.**

STARTING OPTIONS
____ Prerequisite Skills Review: CRB pages 5–6
____ Strategies for Reading Mathematics: CRB pages 7–8
____ Homework Check: TE page 509: Answer Transparencies
____ Warm-Up or Daily Homework Quiz: TE pages 527 and
          513, CRB page 11, or Transparencies

TEACHING OPTIONS
____ Motivating the Lesson: TE page 528
____ Lesson Opener (Geometry Software): CRB page 12 or Transparencies
____ Technology Activity with Keystrokes: CRB page 13
____ Examples 1–3: SE pages 528–530
____ Extra Examples: TE pages 528–530 or Transparencies
____ Closure Question: TE page 530
____ Guided Practice Exercises: SE page 531

APPLY/HOMEWORK
Homework Assignment
____ Block Schedule: 11–30, 33–36, 41, 42, 44–50 even

Reteaching the Lesson
____ Practice Masters: CRB pages 14–16 (Level A, Level B, Level C)
____ Reteaching with Practice: CRB pages 17–18 or Practice Workbook with Examples
____ Personal Student Tutor

Extending the Lesson
____ Applications (Interdisciplinary): CRB page 20
____ Challenge: SE page 534; CRB page 21 or Internet

ASSESSMENT OPTIONS
____ Checkpoint Exercises: TE pages 528–530 or Transparencies
____ Daily Homework Quiz (9.1): TE page 534, CRB page 24, or Transparencies
____ Standardized Test Practice: SE page 534; TE page 534; STP Workbook; Transparencies

CHAPTER PACING GUIDE	
Day	**Lesson**
1	Assess Ch. 8; **9.1 (all)**
2	9.2 (all); 9.3 (begin)
3	9.3 (end); 9.4 (begin)
4	9.4 (end); 9.5 (begin)
5	9.5 (end); 9.6 (all)
6	9.7 (all)
7	Review Ch. 9; Assess Ch. 9

Notes _____

_____

_____

NAME _____ DATE _____

WARM-UP EXERCISES

For use before Lesson 9.1, pages 527–534

Solve the proportion.

1. $\dfrac{x}{4} = \dfrac{12}{16}$ **2.** $\dfrac{4}{y} = \dfrac{y}{9}$ **3.** $\dfrac{12}{6} = \dfrac{6}{r}$

4. $\dfrac{4}{x} = \dfrac{x}{12}$ **5.** $\dfrac{y-2}{6} = \dfrac{6}{2}$

DAILY HOMEWORK QUIZ

For use after Lesson 8.7, pages 506–514

1. Identify the dilation and find its scale factor. Then find the values of the variables.

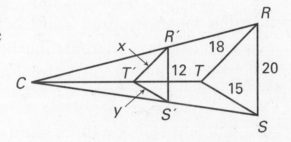

Use the origin as the center of the dilation, the given scale factor, and the given coordinates of the vertices of the preimage polygon to find the coordinates of the vertices of the image polygon.

2. $\triangle XYZ$, with scale factor 5 and vertices $X(3, 1)$, $Y(6, 4)$, and $Z(8, -1)$

3. $\square PQRS$ with scale factor $\dfrac{2}{3}$ and vertices $P(-6, 3)$, $Q(3, 3)$, $R(3, -6)$, and $S(-6, -6)$

Geometry Software Lesson Opener

For use with pages 527–534

Lesson 9.1

Use geometry software to construct, separate, and nest the three similar right triangles that result when the altitude is drawn to the hypotenuse of a right triangle. Before you begin, select the option to keep the preimage displayed.

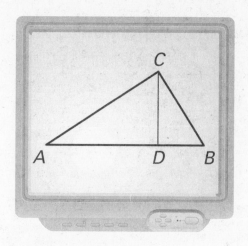

1. Draw a right triangle *ABC* with side $\overline{AB}$ horizontal and right angle at *C*. Measure all the angles of △*ABC*. Construct a perpendicular to $\overline{AB}$ from *C*. Construct *D*, the point of intersection. Hide the perpendicular line and construct $\overline{CD}$. Sketch your result and identify the three similar right triangles: △*CDB*, △*ADC*, and △*ACB*.

2. Now you will separate the three similar triangles, positioning them so that corresponding parts are easy to identify. △*CDB* will stay in place. To separate △*ADC*, mark *A* as a center of rotation and rotate △*ADC* −90° about *A*. To separate △*ABC*, mark $\overline{BC}$ as a mirror line and reflect △*ABC* in $\overline{BC}$. Then mark *C* as a center of rotation and rotate the image of △*ABC* the same measure as ∠*CBA*. Sketch and label each of the three similar right triangles. Write a statement of similarity.

3. Repeat Exercise 1. Now you will nest the three similar triangles, positioning them on top of each other. △*CDB* will stay in place. Mark *D* as a center of rotation and rotate △*ADC* −90° about *D*. Mark $\overline{BC}$ as a mirror line and reflect △*ABC* in $\overline{BC}$. Mark *C* as a center of rotation and rotate the image of △*ABC* the same measure as ∠*CBA*. Translate this image down the length of $\overline{CD}$. Sketch the three nested triangles. Do they appear to be similar?

Technology Activity Keystrokes

For use with page 533

Keystrokes for Exercises 37 and 38

TI-92

1. Draw △ABC so that it is not a right triangle using the triangle command (**F3** 3).

2. Draw altitude $\overline{CD}$ (make sure the altitude is inside △ABC; redraw △ABC if necessary). **F4** 1 (Place cursor on point C.) **ENTER** (Move cursor to $\overline{AB}$.) **ENTER** Label the intersection of the perpendicular line and $\overline{AB}$ point D (**F2** 3).

3. Measure ∠ACB.

   F6 3 (Place cursor on point A.) **ENTER** (Move cursor to point C.) **ENTER** (Move cursor to point B.) **ENTER**

   Measure $\overline{AD}$, $\overline{CD}$, and $\overline{BD}$.

   F6 1 (Place cursor on point A.) **ENTER** (Move cursor to point D.) **ENTER** Repeat this process for the other two sides.

4. Calculate the ratios $\dfrac{BD}{CD}$ and $\dfrac{CD}{AD}$.

   F6 6 (Use cursor to highlight the length of $\overline{BD}$.) **ENTER** **÷** (Highlight the length of CD.) **ENTER** **ENTER** (The result will appear on the screen.) Repeat this process for the second ratio.

5. Drag point C until angle ACB is equal to 90°.

   F1 1 (Place cursor on point C.) **ENTER** (Use the drag key 🖐 and the cursor pad to drag the point.)

SKETCHPAD

1. Draw △ABC so that it is not a right triangle. Select segment from the straightedge tools.

2. Draw altitude $\overline{CD}$ (make sure the altitude is inside △ABC; redraw △ABC if necessary). Use the selection arrow tool to select $\overline{AB}$ and point C. Choose **Perpendicular Line** from the **Construct** menu. Label the intersection of the perpendicular line and $\overline{AB}$ point D.

3. Measure ∠ACB. To measure ∠ACB, use the selection arrow tool to select point A, hold the shift key down, and select point C and B. Then choose **Angle** from the **Measure** menu. Measure $\overline{AD}$, $\overline{CD}$, and $\overline{BD}$. Choose **Length** from the **Measure** menu.

4. Calculate the ratios $\dfrac{BD}{CD}$ and $\dfrac{CD}{AD}$. Choose **Calculate** from the **Measure** menu. Click the length of $\overline{BD}$, click **÷**, click the length of $\overline{CD}$, and click OK. Repeat this process for the other ratio.

5. Drag point C until ∠ACB is equal to 90° using the translate selection arrow tool.

NAME _____ DATE _____

Practice A
For use with pages 527–534

Use the diagrams at the right.

1. Name the legs of △ABC.

2. Name the hypotenuse of △ABC.

3. What is the measure of ∠A and ∠C?

4. Name the legs of △DEF.

5. Name the hypotenuse of △DEF.

6. Could △DEF have an obtuse angle? Explain.

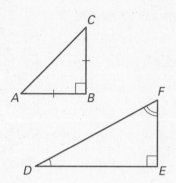

Find the geometric mean of the numbers. Simplify.

7. 16 and 5   8. 9 and 25   9. 6 and 49

Use the diagram.

10. Sketch the three similar triangles in the diagram. Label the vertices

11. Write similarity statements for the three triangles.

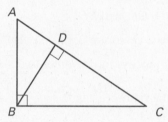

Complete and solve the proportion.

12. $\dfrac{8}{4} = \dfrac{?}{x}$

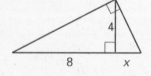

13. $\dfrac{9}{x} = \dfrac{x}{?}$

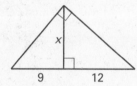

14. $\dfrac{8}{x} = \dfrac{x}{?}$

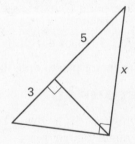

Write similarity statements for the three similar triangles in the diagram. Then complete the proportion.

15. $\dfrac{AD}{CD} = \dfrac{?}{DB}$

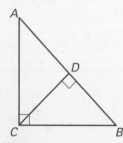

16. $\dfrac{EG}{FG} = \dfrac{FG}{?}$

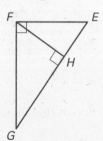

17. $\dfrac{IJ}{IK} = \dfrac{?}{IL}$

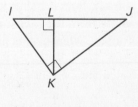

Geometry
Chapter 9 Resource Book

NAME _____ DATE _____

Practice B

For use with pages 527–534

Complete and solve the proportion.

1. $\dfrac{x}{12} = \dfrac{?}{8}$

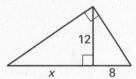

2. $\dfrac{15}{x} = \dfrac{x}{?}$

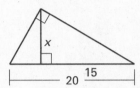

3. $\dfrac{9}{x} = \dfrac{x}{?}$

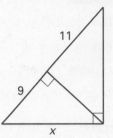

Write similarity statements for three similar triangles in the diagram. Then find the given length.

4. Find QS.

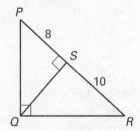

5. Find TU.

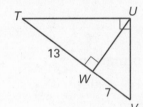

6. Find XZ.

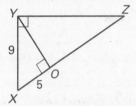

Find the value of each variable.

7.

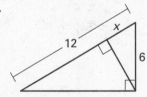

8.

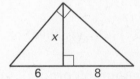

9.

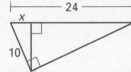

Complete the proof.

10. Given: $\triangle XYZ$ is a right triangle with $m\angle XYZ = 90°$; $\overline{VW} \parallel \overline{XY}$, $\overline{YU}$ is an altitude of $\triangle XYZ$.

Prove: $\triangle YUZ \sim \triangle VWZ$

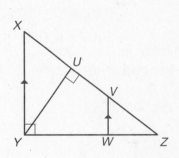

Statements	Reasons
1. $\triangle XYZ$ is a right $\triangle$ with altitude $\overline{YU}$.	1. _____?_____
2. $\triangle XYZ \sim \triangle YUZ$	2. _____?_____
3. $\overline{VW} \parallel \overline{XY}$	3. _____?_____
4. $\angle VWZ \cong \angle XYZ$	4. _____?_____
5. $\angle Z \cong \angle Z$	5. _____?_____
6. $\triangle XYZ \sim \triangle VWZ$	6. _____?_____
7. $\triangle YUZ \sim \triangle VWZ$	7. _____?_____

NAME _____ DATE _____

Practice C

For use with pages 527–534

Use the diagrams at the right to find the indicated length.

1. $AD = 16, DB = 12, DC =$ ___?___

2. $AB = 20, AD = 16, AC =$ ___?___

3. $AD = 16, DC = 2, BC =$ ___?___

4. $DC = 4, BC = 6, AC =$ ___?___

5. $AD = 25, DB = 10, DC =$ ___?___

6. $AD = 4, DC = 1, DB =$ ___?___

In Exercises 7–9, use the diagram of the squat machine where $ZY = 36$ in. and $ZW = 24$ in.

7. Find the length of the vertical support bar, XY.

8. Find the length of the base bar, WX.

9. Find the length of the cross bar, XZ.

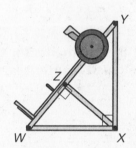

In Exercises 10–14, use the given information.

Given: $\triangle ABC$ is a right triangle with $m\angle C = 90°$,

$\overline{DC} \perp \overline{AB}, \overline{FD}$ bisects $\angle ADC, \overline{ED}$ bisects $\angle BDC$

10. Which angles are congruent?

11. Which triangles are similar?

12. *True or False?* $\dfrac{AD}{CD} = \dfrac{AC}{BC}$ 13. Is $\overline{DF}$ an altitude of $\triangle ADC$?   14. *True or False?* $\dfrac{CE}{CB} = \dfrac{CF}{CA}$

Write a two-column proof or a paragraph proof.

15. **Given:** $\triangle ABC$ with altitude $\overline{BD}$,

$\quad\quad m\angle ABC = 90°$,

$\quad\quad AC = 6, DC = 4$

Prove: $BC = 2\sqrt{6}$

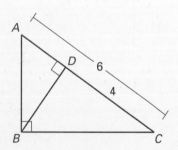

16. **Given:** $\triangle JKL$ with altitude $\overline{KM}$,

$\quad\quad m\angle LKJ = 90°$,

$\quad\quad KM = 3, KJ = 5$

Prove: $JL = \dfrac{25}{4}$

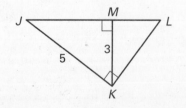

Geometry
Chapter 9 Resource Book

Reteaching with Practice

For use with pages 527–534

GOAL Solve problems involving similar right triangles formed by the altitude drawn to the hypotenuse of a right triangle and use a geometric mean to solve problems

---

Theorem 9.1
If the altitude is drawn to the hypotenuse of a right triangle, then the two triangles formed are similar to the original triangle and to each other.

Theorem 9.2
In a right triangle, the altitude from the right angle to the hypotenuse divides the hypotenuse into two segments. The length of the altitude is the geometric mean of the lengths of the two segments.

Theorem 9.3
In a right triangle, the altitude from the right angle to the hypotenuse divides the hypotenuse into two segments. The length of each leg of the right triangle is a geometric mean of the lengths of the hypotenuse and the segment of the hypotenuse that is adjacent to the leg.

EXAMPLE 1 *Finding the Height of a Triangle*

Consider the right triangle shown.

a. Identify the similar triangles.

b. Find the height h of $\triangle ABC$.

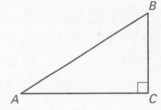

SOLUTION

a. $\triangle ABC \sim \triangle CBD \sim \triangle ACD$
Sketch the three similar right triangles so that the corresponding angles and sides have the same orientation.

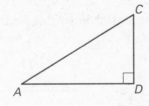

b. Use the fact that $\triangle ABC \sim \triangle CBD$ to write a proportion.

$$\frac{CD}{AC} = \frac{CB}{AB}$$ Corresponding side lengths are in proportion.

$$\frac{h}{10} = \frac{6}{11.6}$$ Substitute.

$$11.6h = 6(10)$$ Cross product property

$$h \approx 5.2$$ Solve for h.

NAME _____ DATE _____

Reteaching with Practice

For use with pages 527–534

Exercises for Example 1

Find the height, *h*, of the given right triangle.

1.
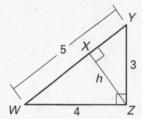

2.
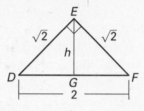

3.
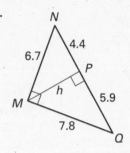

EXAMPLE 2 *Using a Geometric Mean*

Find the value of each variable.

a.
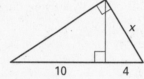

b.
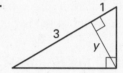

SOLUTION

a. Apply Theorem 9.3.

$$\frac{10 + 4}{x} = \frac{x}{4}$$

$$x^2 = 56$$

$$x = \sqrt{56} = \sqrt{4 \cdot 14} = 2\sqrt{14}$$

b. Apply Theorem 9.2.

$$\frac{3}{y} = \frac{y}{1}$$

$$y^2 = 3$$

$$y = \sqrt{3}$$

Exercises for Example 2

Find the value of each variable to the nearest tenth.

4.
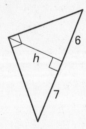

5.
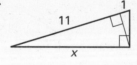

6.
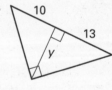

NAME _____ DATE _____

Quick Catch-Up for Absent Students

For use with pages 527–534

The items checked below were covered in class on (date missed) _____

Lesson 9.1: Similar Right Triangles

_____ **Goal 1:** Solve problems involving similar right triangles formed by the altitude drawn to the hypotenuse of a right triangle. (pp. 527–528)

Material Covered:

_____ Activity: Investigating Similar Right Triangles

_____ Student Help: Study Tip

_____ Example 1: Finding the Height of a Roof

_____ **Goal 2:** Use a geometric mean to solve problems. (pp. 529–530)

Material Covered:

_____ Student Help: Look Back

_____ Student Help: Skills Review

_____ Student Help: Study Tip

_____ Example 2: Using a Geometric Mean

_____ Example 3: Using Indirect Measurement

_____ Other (specify) _____

_____

Homework and Additional Learning Support

_____ Textbook (specify) _pp. 531–534_____

_____

_____ *Reteaching with Practice* worksheet (specify exercises)_____

_____ *Personal Student Tutor* for Lesson 9.1

Interdisciplinary Application

For use with pages 527–534

Nautilus

BIOLOGY   Nautilus are a type of mollusk called a cephalopod. Although the squid and the octopus are also cephalopods, the nautilus has the unique characteristic of inhabiting a shell. The shell of a chambered nautilus has a maximum diameter of approximately 10 inches and includes up to 36 chambers. The nautilus lives in the largest, most recent chamber where its 90 to 94 tentacles can easily reach out to capture food like crab and shrimp. The only connection between the nautilus and its earlier chambers is a tube called the siphuncle, which allows the mollusk to control its vertical movement.

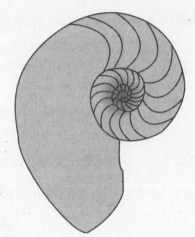

A sliced nautilus shell, as shown at the right, is an excellent example of the relationship between nature and mathematics. The spiral formed by the chambers of the shell can be modeled by an equiangular spiral.

In Exercises 1–3, use the diagram at the right.

1. Write a proportion relating the given segments.
   a. $\overline{OU}, \overline{OV},$ and $\overline{OW}$ b. $\overline{OV}, \overline{OW},$ and $\overline{OX}$
   c. $\overline{OW}, \overline{OX},$ and $\overline{OY}$ d. $\overline{OX}, \overline{OY},$ and $\overline{OZ}$

2. Complete the statement: ___?___ is the geometric mean of $\overline{OU}$ and $\overline{OW}$.

3. Use your answers from Exercises 1 and 2 to describe a pattern in the sequence of segments.

4. Use Theorem 9.3 to write two proportions that hold for $\triangle XYZ$.

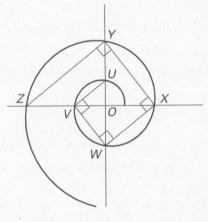

NAME _____ DATE _____

Challenge: Skills and Applications

For use with pages 527–534

In Exercises 1 and 2, use the following information.

A triangle inscribed in a circle is a right triangle if and only if the longest side of the triangle is a diameter of the circle.

1. Given point D on line segment $\overline{AB}$, explain how to use a compass and straightedge to construct a line segment whose length is the geometric mean of AD and BD.

2. Refer to the diagram. In $\triangle ABC$, $\overline{CD}$ is an altitude and $\overline{CE}$ is a median.

   a. Explain why CE is the arithmetic mean of AD and BD.

   b. Use the diagram to show that the arithmetic mean of AD and BD is greater than the geometric mean of AD and BD.

   c. Use your argument from part (b) to show that the arithmetic mean of any two distinct positive numbers is greater than the geometric mean.

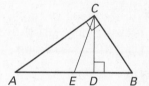

In Exercises 3 and 4, refer to the diagram.

3. Prove that $\dfrac{(AC)^2}{(BC)^2} = \dfrac{AD}{BD}$.

4. If $AD = x^2$ and $BD = y^2$, use the Geometric Mean Theorems to find AC, BC, and CD in terms of x and y. (Assume that x and y are positive.)

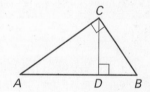

In Exercises 5–10, find the possible values of x.

5.
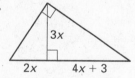

6.
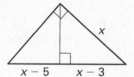

7.
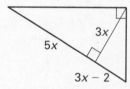

8.
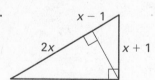

9.
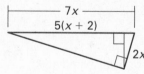

10.
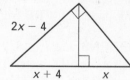

TEACHER'S NAME _____ CLASS _____ ROOM _____ DATE _____

Lesson Plan

1-day lesson (See *Pacing the Chapter,* TE pages 524C–524D)   **For use with pages 535–541**

GOALS   1. **Prove the Pythagorean Theorem.**
2. **Use the Pythagorean Theorem to solve real-life problems.**

State/Local Objectives _____

_____

✓ Check the items you wish to use for this lesson.

STARTING OPTIONS
_____ Homework Check: TE page 531: Answer Transparencies
_____ Warm-Up or Daily Homework Quiz: TE pages 535 and 534, CRB page 24, or Transparencies

TEACHING OPTIONS
_____ Motivating the Lesson: TE page 536
_____ Lesson Opener (Activity): CRB page 25 or Transparencies
_____ Technology Activity with Keystrokes: CRB pages 26–28
_____ Examples 1–4: SE pages 536–537
_____ Extra Examples: TE pages 536–537 or Transparencies
_____ Closure Question: TE page 537
_____ Guided Practice Exercises: SE page 538

APPLY/HOMEWORK
Homework Assignment
_____ Basic   8–24 even, 25–31, 34, 39, 42–56 even
_____ Average   8–24 even, 25–31, 33–36, 39, 42–56 even
_____ Advanced   8–24 even, 25–31, 33–41, 42–56 even

Reteaching the Lesson
_____ Practice Masters: CRB pages 29–31 (Level A, Level B, Level C)
_____ Reteaching with Practice: CRB pages 32–33 or Practice Workbook with Examples
_____ Personal Student Tutor

Extending the Lesson
_____ Applications (Real-Life): CRB page 35
_____ Challenge: SE page 541; CRB page 36 or Internet

ASSESSMENT OPTIONS
_____ Checkpoint Exercises: TE pages 536–537 or Transparencies
_____ Daily Homework Quiz (9.2): TE page 541, CRB page 39, or Transparencies
_____ Standardized Test Practice: SE page 541; TE page 541; STP Workbook; Transparencies

Notes _____

_____

_____

Lesson 9.2

TEACHER'S NAME _____ CLASS _____ ROOM _____ DATE _____

Lesson Plan for Block Scheduling

Half-day lesson (See *Pacing the Chapter,* TE pages 524C–524D)   For use with pages 535–541

GOALS   1. Prove the Pythagorean Theorem.
2. Use the Pythagorean Theorem to solve real-life problems.

State/Local Objectives _____

_____

_____

✓ **Check the items you wish to use for this lesson.**

STARTING OPTIONS

____ Homework Check: TE page 531: Answer Transparencies
____ Warm-Up or Daily Homework Quiz: TE pages 535 and
        534, CRB page 24, or Transparencies

TEACHING OPTIONS

____ Motivating the Lesson: TE page 536
____ Lesson Opener (Activity): CRB page 25 or Transparencies
____ Technology Activity with Keystrokes: CRB pages 26–28
____ Examples 1–4: SE pages 536–537
____ Extra Examples: TE pages 536–537 or Transparencies
____ Closure Question: TE page 537
____ Guided Practice Exercises: SE page 538

APPLY/HOMEWORK
Homework Assignment (See also the assignment for Lesson 9.3.)
____ Block Schedule: 8–24 even, 25–31, 33–36, 39, 42–56 even

Reteaching the Lesson

____ Practice Masters: CRB pages 29–31 (Level A, Level B, Level C)
____ Reteaching with Practice: CRB pages 32–33 or Practice Workbook with Examples
____ Personal Student Tutor

Extending the Lesson

____ Applications (Real-Life): CRB page 35
____ Challenge: SE page 541; CRB page 36 or Internet

ASSESSMENT OPTIONS

____ Checkpoint Exercises: TE pages 536–537 or Transparencies
____ Daily Homework Quiz (9.2): TE page 541, CRB page 39, or Transparencies
____ Standardized Test Practice: SE page 541; TE page 541; STP Workbook; Transparencies

CHAPTER PACING GUIDE	
Day	Lesson
1	Assess Ch. 8; 9.1 (all)
2	**9.2 (all)**; 9.3 (begin)
3	9.3 (end); 9.4 (begin)
4	9.4 (end); 9.5 (begin)
5	9.5 (end); 9.6 (all)
6	9.7 (all)
7	Review Ch. 9; Assess Ch. 9

Lesson 9.2

Notes _____

_____

_____

WARM-UP EXERCISES

For use before Lesson 9.2, pages 535–541

Solve the equation. Express the answer to the nearest tenth.

1. $\dfrac{8}{c} = \dfrac{c}{9}$ **2.** $c^2 = 36$ **3.** $c^2 = 84.2$

4. $c^2 = 28.2 + 42.1$ **5.** $c^2 - 81 = 144$

DAILY HOMEWORK QUIZ

For use after Lesson 9.1, pages 527–534

Use the diagram.

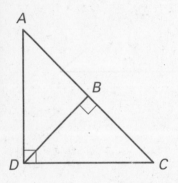

1. Write similarity statements for the three triangles.

2. Complete the proportion.

$$\frac{AD}{DC} = \frac{CB}{?}$$

Find the value of each variable.

3.

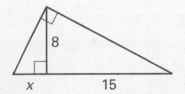

SET UP: Work with a partner.

Half of a proof of the Pythagorean Theorem is shown. Below that, the remaining five statements of the proof are given out of order. Complete the proof by writing the other statements in order with a reason for each.

Pythagorean Theorem: In a right triangle, the square of the length of the hypotenuse is equal to the sum of the squares of the lengths of the legs.

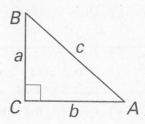

Given: In $\triangle ABC$, $\angle BCA$ is a right angle.

Prove: $a^2 + b^2 = c^2$

1. Extend $\overline{AC}$ to D such that $\angle DBA$ is a right angle.

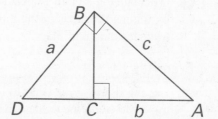

2. $\dfrac{b}{a} = \dfrac{a}{DC}$ (Geometric Mean Theorem)

3. $DC = \dfrac{a^2}{b}$ (Cross product prop.; multiplication prop. of equality)

4. $\triangle BCD \sim \triangle ABD$ (An altitude to the hyp. of a rt. $\triangle$ divides the $\triangle$ into 2 similar $\triangle$'s whic triangles also $\sim$ to the original $\triangle$.)

5. $\dfrac{DC}{DB} = \dfrac{BC}{BA}$ (Ratios of lengths of corresponding sides of similar triangles are equal.)

Area of $\triangle BCD$ + Area of $\triangle BCA$ = Area of $\triangle DBA$	
$DB = \dfrac{ac}{b}$	$a^2 + b^2 = c^2$
$\dfrac{1}{2} \cdot a \cdot \dfrac{a^2}{b} + \dfrac{1}{2} \cdot a \cdot b = \dfrac{1}{2} \cdot \dfrac{ac}{b} \cdot c$	$\dfrac{\frac{a^2}{b}}{DB} = \dfrac{a}{c}$

NAME _____ DATE _____

Technology Activity

For use with pages 535–541

GOAL **To verify a proof of the Pythagorean Theorem using geometry software**

There are many different proofs of the Pythagorean Theorem. Former United States President, James Garfield, wrote a proof of the Pythagorean Theorem in 1836.

ACTIVITY

1 Draw trapezoid *PQRS* with vertices $(-1, 0)$, $(-1, 4)$, $(6, 3)$, and $(6, 0)$.

2 Draw point *T* at $(2, 0)$. Draw $\overline{QT}$ and $\overline{RT}$. This divides trapezoid *PQRS* into three right triangles.

3 Calculate the area of trapezoid *PQRS* using the formula $A = \frac{1}{2}h(b_1 + b_2)$.

4 Calculate the area of trapezoid *PQRS* by finding the areas of the three right triangles drawn in Step 2.

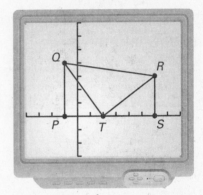

Exercises

1. Use the figure at the right to find the area of the trapezoid (with the values given below) two different ways. (*Hint:* Use the formula for the area of a trapezoid and find the area of the three triangles and add them together.)

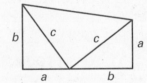

   a. If $a = 3$, $b = 4$, and $c = 5$

   b. If $a = 5$, $b = 12$, and $c = 13$

   c. If $a = 6$, $b = 7$, and $c = \sqrt{85}$

2. Use the figure from Exercise 1 to construct Garfield's proof algebraically by finding the area of the trapezoid two different ways and setting them equal to each other.

LESSON
9.2
CONTINUED

NAME _____ DATE _____

Technology Activity Keystrokes

For use with pages 535–541

TI-92

1. Turn on the axes and the grid.

   F8 9 (Set Coordinate Axes to RECTANGULAR and Grid to ON.) **ENTER** Draw
   trapezoid *PQRS* with vertices $(-1, 0)$, $(-1, 4)$, $(6, 3)$, and $(6, 0)$. **F3** 4 (Move
   cursor to $(-1, 0)$ and prompt says "POINT ON . . .") **ENTER** 2 (Move cursor to
   point $(-1, 4)$.) **ENTER** (Move cursor point to $(6, 3)$.) **ENTER** (Move cursor to
   $(6, 0)$ and prompt says "POINT ON. . .") **ENTER** 2 (Move cursor to $(-1, 0)$.)
   ENTER

   Label the vertices.

   F7 4 (Move cursor to vertex at $(-1, 0)$.) **ENTER** *P* **ENTER** (Move cursor to
   vertex at $(-1, 4)$.) **ENTER** *Q* **ENTER** (Move cursor to vertex at $(6, 3)$.) **ENTER**
   R **ENTER** (Move cursor to vertex at $(6, 0)$.) **ENTER** *S* **ENTER**

2. Draw point *T* at $(2, 0)$.

   F2 1 (Move cursor to side $(2, 0)$ and prompt says "POINT ON . . .")

   ENTER 2*T*

   Draw $\overline{QT}$ and $\overline{RT}$ using the segment command (**F2** 5).

3. Use the distance and length command (**F6** 1) to find the height of
   trapezoid *PQRS* and the length of both of its bases. Then use the calculate
   command (**F6** 6) to enter the formula and find the area.

4. Use the distance and length command (**F6** 1) to find the height of each
   triangle and the length of the base of each triangle. Then use the calculate
   command (**F6** 6) to enter the formula and find the area of each triangle.

Technology Activity Keystrokes

For use with pages 535–541

SKETCHPAD

1. Turn on the axes and the grid. Choose **Snap To Grid** from the **Graph** menu. Use the segment straightedge tool to draw trapezoid *PQRS* with vertices $P(-1, 0)$, $Q(-1, 4)$, $R(6, 3)$, and $S(6, 0)$. Use the text tool to label the points.

2. Use the point tool to draw point *T* at $(2, 0)$. Use the text tool to label the point. Use the segment straightedge tool to draw $\overline{QT}$ and $\overline{RT}$.

3. Find the height of trapezoid *PQRS* and the length of its bases. Then choose **Calculate** from the **Measure** menu and enter the area formula for a trapezoid using the appropriate values.

4. Find the height of each triangle and the length of each base. Then choose **Calculate** from the **Measure** menu and enter the area formula for a triangle using the appropriate values for each triangle.

Geometry
Chapter 9 Resource Book

Practice A

For use with pages 535–541

Use the labeled triangles to state the Pythagorean Theorem.

1.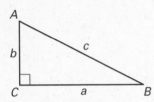

2.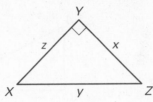

3.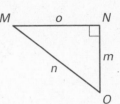

Simplify the radical.

4. $\sqrt{12}$

5. $\sqrt{48}$

6. $\sqrt{20}$

7. $\sqrt{18}$

8. $\sqrt{60}$

9. $\sqrt{75}$

Find the unknown side length. Simplify answers that are radicals. Tell whether the side lengths form a Pythagorean triple.

10.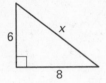

11.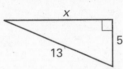

12.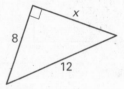

13.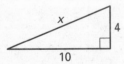

14.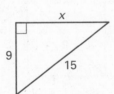

15.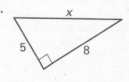

Find the area of the figure. Round decimal answers to the nearest tenth.

16.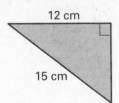

17.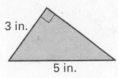

18.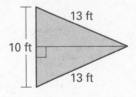

Solve. Round your answer to the nearest tenth.

19. A 48-inch wide screen television means that the measure along the diagonal is 48 inches. If the screen is a square, what are the dimensions of the length and width?

20. The doorway of the family room measures $6\frac{1}{2}$ feet by 3 feet. What is the length of the diagonal of the doorway?

21. You place a 10-foot ladder against a wall. If the base of the ladder is 3 feet from the wall, how high up the wall does the top of the ladder reach?

Lesson 9.2

NAME _____ DATE _____

Practice B

For use with pages 535–541

Use △**ABC** to determine if the equation is *true* or *false*.

1. $b^2 + a^2 = c^2$

2. $c^2 - a^2 = b^2$

3. $b^2 - c^2 = a^2$

4. $c^2 = a^2 - b^2$

5. $c^2 = b^2 + a^2$

6. $a^2 = c^2 - b^2$

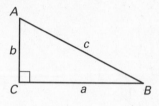

Find the unknown side length. Simplify answers that are radicals. Tell whether the side lengths form a Pythagorean triple.

7.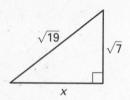

8.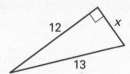

9.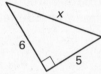

10.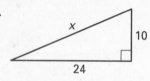

11.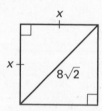

12.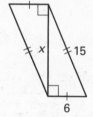

Find the area of the figure. Round decimal answers to the nearest tenth.

13.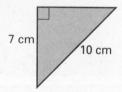

14.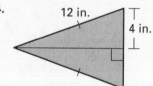

15.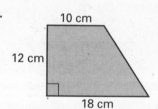

Solve. Round your answer to the nearest tenth.

16. A smaller commuter airline flies to three cities whose locations form the vertices of a right triangle. The total flight distance (from city A to city B to city C and back to city A) is 1400 miles. It is 600 miles between the two cities that are furthest apart. Find the other two distances between cities.

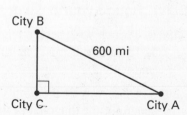

17. Each base on a standard baseball diamond lies 90 feet from the next. Find the distance the catcher must throw a baseball from 3 feet behind home plate to second base.

NAME _____ DATE _____

Practice C
For use with pages 535–541

Find the unknown side length. Simplify answers that are radicals. Tell whether the side lengths form a Pythagorean triple.

1.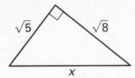

2.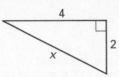

3.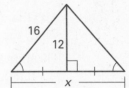

4.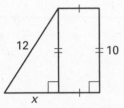

5.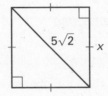

6.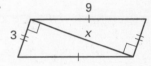

Find the missing length so that x, y, and z are Pythagorean triples.

7. $x = 6, y = 8$

8. $y = 24, z = 26$

9. $x = 16, y = 30$

10. $x = 24, z = 51$

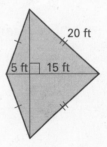

Find the area of the figure. Round decimal answers to the nearest tenth.

11.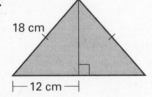

18 cm

12 cm

12.

6 in.

9 in.

15 in.

13.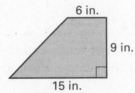

20 ft

5 ft   15 ft

14. A standard doorway measures 6 feet 8 inches by 3 feet. What is the largest dimension that will fit through the doorway without bending?

15. Use the Pythagorean Theorem and the diagram at the right to show $AB = \sqrt{(x_2 - x_1)^2 + (y_2 - y_1)^2}$. That is, show the distance formula is true.

16. Solve for x in the partial spiral shown at the right.

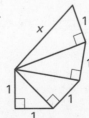

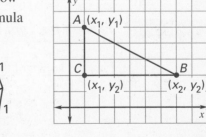

NAME _____ DATE _____

Reteaching with Practice

For use with pages 535–541

GOAL **Use the Pythagorean Theorem to solve problems**

> ### VOCABULARY
>
> A **Pythagorean triple** is a set of three positive integers a, b, and c, that satisfy the equation $c^2 = a^2 + b^2$.
>
> **Theorem 9.4   Pythagorean Theorem**
> In a right triangle, the square of the length of the hypotenuse is equal to the sum of the squares of the lengths of the legs.

EXAMPLE 1 *Find the Length of a Hypotenuse*

Find the length of the hypotenuse of the right triangle.

Tell whether the side lengths form a Pythagorean triple.

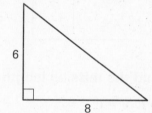

SOLUTION

$(\text{hypotenuse})^2 = (\text{leg})^2 + (\text{leg})^2$	Pythagorean Theorem
$x^2 = 6^2 + 8^2$	Substitute.
$x^2 = 36 + 64$	Multiply.
$x^2 = 100$	Add.
$x = 10$	Find the positive square root.

Because the side lengths 6, 8, and 10 are integers, they form a Pythagorean triple.

Exercises for Example 1
....

Find the length of the hypotenuse of the right triangle. Tell whether the side lengths form a Pythagorean triple.

1.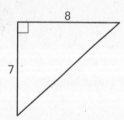

2.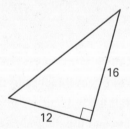

3.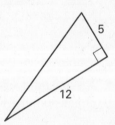

EXAMPLE 2 *Finding the Length of a Leg*

Find the length of the leg of the right triangle.

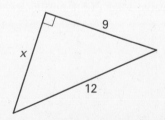

NAME _____ DATE _____

Reteaching with Practice

For use with pages 535–541

SOLUTION

$$(\text{hypotenuse})^2 = (\text{leg})^2 + (\text{leg})^2 \qquad \text{Pythagorean Theorem}$$

$$12^2 = 9^2 + x^2 \qquad \text{Substitute.}$$

$$144 = 81 + x^2 \qquad \text{Multiply.}$$

$$63 = x^2 \qquad \text{Subtract 81 from each side.}$$

$$\sqrt{63} = x \qquad \text{Find the positive square root.}$$

Exercises for Example 2

Find the unknown side length. Round to the nearest tenth, if necessary.

4.
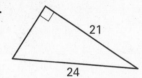
21
24

5.
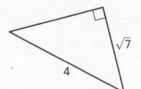
$\sqrt{7}$
4

6.
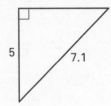
5
7.1

EXAMPLE 3 *Finding the Area of a Triangle*

Find the area of the triangle to the nearest tenth.

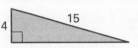
4   15

SOLUTION

In this case, the side of length 4 can be used as the height and the side of unknown length can be used as the base. To find the length of the unknown side, use the Pythagorean Theorem.

$$(\text{hypotenuse})^2 = (\text{leg})^2 + (\text{leg})^2 \qquad \text{Pythagorean Theorem}$$

$$15^2 = 4^2 + b^2 \qquad \text{Substitute.}$$

$$\sqrt{209} = b \qquad \text{Solve for } b.$$

Now find the area of the triangle.

$$A = \frac{1}{2}bh = \frac{1}{2}\left(\sqrt{209}\right)(4) \approx 28.9 \text{ square units}$$

Exercises for Example 3

Find the area of the triangle to the nearest tenth.

7.
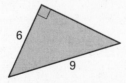
6
9

8.
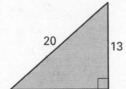
20   13

9.
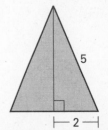
5
├─ 2 ─┤

Lesson 9.2

NAME _____ DATE _____

Quick Catch-Up for Absent Students

For use with pages 535–541

The items checked below were covered in class on (date missed) _____

Lesson 9.2: The Pythagorean Theorem

_____ **Goal 1:** Prove the Pythagorean Theorem. (p. 535)

_____ **Goal 2:** Use the Pythagorean Theorem to solve real-life problems. (pp. 536–537)

Material Covered:

_____ Example 1: Finding the Length of a Hypotenuse

_____ Student Help: Skills Review

_____ Example 2: Finding the Length of a Leg

_____ Student Help: Look Back

_____ Example 3: Finding the Area of a Triangle

_____ Example 4: Indirect Measurement

Vocabulary:

Pythagorean triple, p. 536

_____ Other (specify) _____

_____

Homework and Additional Learning Support

_____ Textbook (specify) <u>pp. 538–541</u> _____

_____

_____ *Reteaching with Practice* worksheet (specify exercises)_____

_____ *Personal Student Tutor* for Lesson 9.2

NAME _____ DATE _____

Real-Life Application: When Will I Ever Use This?

For use with pages 535–541

Baseball

Baseball is one of the most well-known sports in the world, for it is watched and played by millions on almost every continent. With organized teams that range from little leagues to professional leagues, baseball is not limited by age or skill level.

This historical background of this popular summertime pastime is uniquely American. Although baseball may be similar to an English game called rounders, most agree that Abner Doubleday invented the modern form in Cooperstown, NY in 1839. The National Baseball Hall of Fame is located in Cooperstown to commemorate the event.

In Exercises 1–5, use the following information and the diagram below.

José plays second base for his high school baseball team. The opposing team has a player on first who is known for his running speed and base stealing. The scouting report states that this player can run a five second 40-yard dash. When the pitcher releases the ball, the runner on first heads toward second base. By the time the catcher from José's team catches the ball and then throws it towards second, the runner is half-way to second base. The catcher throws the ball 70 miles per hour. Will José be able to tag the runner out at second?

1. Compute the speed of the runner in feet per second.

2. Compute the speed of the throw in feet per second.

3. Find the time it takes the runner to reach second. (*Hint:* distance = rate × time)

4. Find the time it takes the throw to reach second.

5. Can José tag the runner out?

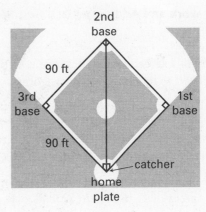

NAME _____ DATE _____

Challenge: Skills and Applications

For use with pages 535–541

In Exercises 1–4, *WXYZ* is the square base of a pyramid, and segments $\overline{WY}$ and $\overline{XZ}$ intersect at *U*. Also, $\overline{VU}$ is perpendicular to the plane containing *WXYZ*, and $\overline{VT}$ is an altitude of $\triangle VWZ$.

1. If *TU* = 20 and *UV* = 17, find *VW*.

2. If *YZ* = 16 and *VW* = 18, find *UV*.

3. If *UV* = 7 and *VW* = 11, find *TU* and *TV*.

4. If *UV* = 31 and *VW* = 67, find the area of *WXYZ*.

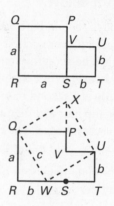

5. Here is a proof of the Pythagorean Theorem.

   a. Given *a* and *b* with $a \geq b$, let *PQRS* be a square with side length *a*, and let *STUV* be a square with side length *b*, as shown. What is the total area of the two squares?

   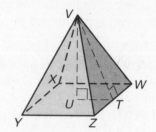

   b. Choose point *W* on $\overline{RS}$ such that *RW* = *b*, and choose point *X* on $\overrightarrow{VP}$ so that *PX* = *b*. Sketch auxiliary lines as shown. Write a paragraph proof that $\triangle QPX \cong \triangle QRW \cong \triangle WTU \cong \triangle XVU$.

   c. Let *c* = *QW*. Explain why $a^2 + b^2 = c^2$.

6. In this exercise, you will prove a theorem that is related to the Pythagorean Theorem.

   Given: $\overline{OA} \perp \overline{OB}, \overline{OB} \perp \overline{OC}, \overline{OA} \perp \overline{OC}$, *OA* = *a*, *OB* = *b*, and *OC* = *c*. Complete the following steps to show that (area of $\triangle ABC)^2$ is equal to the sum of the squares of the areas of $\triangle ABO$, $\triangle BCO$, and $\triangle ACO$.

   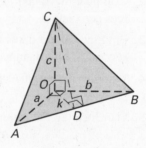

   a. Write an expression for the sum of the squares of the areas of $\triangle ABO$, $\triangle BCO$, and $\triangle ACO$ in terms of *a*, *b*, and *c*.

   b. Choose *D* on $\overline{AB}$ so that $\overline{OD}$ is an altitude of $\triangle ABO$. Let *k* = *OD*. Find an expression for (area of $\triangle ABC)^2$ in terms of *a*, *b*, *c*, and *k*.

   c. Write two different expressions for the area of $\triangle ABO$. Equate these expressions to find an equation that relates *k*, *a*, and *b*. Then show that the expressions you wrote in parts (a) and (b) are equal.

LESSON 9.3

Lesson Plan

2-day lesson (See *Pacing the Chapter,* TE pages 524C–524D)   For use with pages 542–549

GOALS
1. Use the Converse of the Pythagorean Theorem to solve problems.
2. Use side lengths to classify triangles by their angle measures.

State/Local Objectives _____

_____

✓ Check the items you wish to use for this lesson.

STARTING OPTIONS
_____ Homework Check: TE page 538: Answer Transparencies
_____ Warm-Up or Daily Homework Quiz: TE pages 543 and 541, CRB page 39, or Transparencies

TEACHING OPTIONS
_____ Motivating the Lesson: TE page 544
_____ Lesson Opener (Application): CRB page 40 or Transparencies
_____ Technology Activity with Keystrokes: CRB pages 41–43
_____ Examples: Day 1: 1–3: SE pages 543–545; Day 2: See the Extra Examples.
_____ Extra Examples: Day 1 or Day 2: 1–3, TE pages 544–545 or Transp.
_____ Technology Activity: SE page 542
_____ Closure Question: TE page 545
_____ Guided Practice: SE page 545   Day 1: Exs. 1–7; Day 2: See Checkpoint Exs. TE pages 544–545

APPLY/HOMEWORK
Homework Assignment
_____ Basic   Day 1: 8–28 even, 32–36 even; Day 2: 9–27 odd, 29, 30, 31–37 odd, 41, 44, 45, 47–57 odd;
         Quiz 1: 1–8
_____ Average   Day 1: 8–28 even, 32–36 even; Day 2: 9–27 odd, 29, 30, 31–37 odd, 40, 41, 44, 45,
         47–57 odd; Quiz 1: 1–8
_____ Advanced   Day 1: 8–28 even, 32–36 even; Day 2: 9–27 odd, 29, 30, 31–37 odd, 40–46, 47–57 odd;
         Quiz 1: 1–8

Reteaching the Lesson
_____ Practice Masters: CRB pages 44–46 (Level A, Level B, Level C)
_____ Reteaching with Practice: CRB pages 47–48 or Practice Workbook with Examples
_____ Personal Student Tutor

Extending the Lesson
_____ Applications (Interdisciplinary): CRB page 50
_____ Challenge: SE page 548; CRB page 51 or Internet

ASSESSMENT OPTIONS
_____ Checkpoint Exercises: Day 1 or Day 2: TE pages 544–545 or Transp.
_____ Daily Homework Quiz (9.3): TE page 549, CRB page 55, or Transparencies
_____ Standardized Test Practice: SE page 548; TE page 549; STP Workbook; Transparencies
_____ Quiz (9.1–9.3): SE page 549; CRB page 52

Notes _____

_____

_____

Lesson 9.3

TEACHER'S NAME _____ CLASS _____ ROOM _____ DATE _____

Lesson Plan for Block Scheduling

1-day lesson (See *Pacing the Chapter*, TE pages 524C–524D)   For use with pages 542–549

GOALS   1. **Use the Converse of the Pythagorean Theorem to solve problems.**
2. **Use side lengths to classify triangles by their angle measures.**

State/Local Objectives _____

_____

✓ **Check the items you wish to use for this lesson.**

CHAPTER PACING GUIDE	
Day	**Lesson**
1	Assess Ch. 8; 9.1 (all)
2	9.2 (all); **9.3 (begin)**
3	**9.3 (end)**; 9.4 (begin)
4	9.4 (end); 9.5 (begin)
5	9.5 (end); 9.6 (all)
6	9.7 (all)
7	Review Ch. 9; Assess Ch. 9

STARTING OPTIONS
_____ Homework Check: TE page 538: Answer Transparencies
_____ Warm-Up or Daily Homework Quiz: TE pages 543 and
      541, CRB page 39, or Transparencies

TEACHING OPTIONS
_____ Motivating the Lesson: TE page 544
_____ Lesson Opener (Application): CRB page 40 or Transparencies
_____ Technology Activity with Keystrokes: CRB pages 41–43
_____ Examples: Day 2: 1–3: SE pages 543–545; Day 3: See the Extra Examples.
_____ Extra Examples: Day 2 or Day 3: 1–3, TE pages 544–545 or Transp.
_____ Technology Activity: SE page 542
_____ Closure Question: TE page 545
_____ Guided Practice: SE page 545   Day 2: Exs. 1–7; Day 3: See Checkpoint Exs. TE pages 544–545

APPLY/HOMEWORK
Homework Assignment (See also the assignments for Lessons 9.2 and 9.4.)
_____ Block Schedule: Day 2: 8–28 even, 32–36 even; Day 3: 9–27 odd, 29, 30, 31–37 odd, 40, 41, 44,
      45, 47–57 odd; Quiz 1: 1–8

Reteaching the Lesson
_____ Practice Masters: CRB pages 44–46 (Level A, Level B, Level C)
_____ Reteaching with Practice: CRB pages 47–48 or Practice Workbook with Examples
_____ Personal Student Tutor

Extending the Lesson
_____ Applications (Interdisciplinary): CRB page 50
_____ Challenge: SE page 548; CRB page 51 or Internet

ASSESSMENT OPTIONS
_____ Checkpoint Exercises: Day 2 or Day 3: TE pages 544–545 or Transp.
_____ Daily Homework Quiz (9.3): TE page 549, CRB page 55, or Transparencies
_____ Standardized Test Practice: SE page 548; TE page 549; STP Workbook; Transparencies
_____ Quiz (9.1–9.3): SE page 549; CRB page 52

Notes _____

_____

_____

Lesson 9.3

NAME _____ DATE _____

WARM-UP EXERCISES

For use before Lesson 9.3, pages 542–549

Solve the equation for the missing variable. Assume all variables are positive. Express the answer to the nearest tenth.

1. $c^2 = 6^2 + 8^2$

2. $c^2 - 4^2 = 3^2$

3. $c^2 - 10^2 = 16^2$

4. $a^2 + 6^2 = 14^2$

5. $(14\sqrt{2})^2 = 14^2 + b^2$

DAILY HOMEWORK QUIZ

For use after Lesson 9.2, pages 535–541

Use the figure.

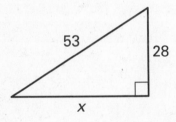

1. Find the unknown side length.

2. Tell whether the side lengths form a Pythagorean triple.

Find the area of the figure.

3.

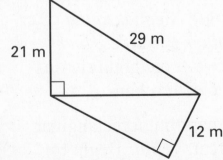

Application Lesson Opener

Geometry is a compound of two Greek words meaning "earth" and "measure." The origin of geometry is generally credited to the ancient Egyptians in their land surveying. Ancient Egyptians would "measure the earth" every year after flooding of the Nile river in order to determine taxes and re-establish property lines. The task of surveying was performed by specialists whose main tool was a rope with knots or marks at equal intervals.

Today we have many sophisticated tools for land surveying. But we also use measuring tapes in the same way that ropes were used thousands of years ago. Here is a common method for measuring a rectangular area of land, using two measuring tapes.

> Measure a base line 4 units long. Mark the endpoints *A* and *B*. Extend one measuring tape from *A* and another measuring tape from *B* to intersect at *C*. Adjust the two tapes until the tape from *B* reads 3 units and the tape from *A* reads 5 units. Mark point *C*. A squarecorner has been created at *B*. Now extend one tape from *A* and another from *B* to intersect at *D* on the same side of $\overline{AB}$ as point *C*. Adjust the two tapes until the tape from *A* reads 3 units and the tape from *B* reads 5 units. Mark point *D*. A square corner has been created at *A*. *ABCD* is a rectangular area that measures 3 units by 4 units. Measure *AC* and *BD* to check for equal diagonals.

1. Draw some diagrams that show the steps of the 3-4-5 method.

2. The Converse of the Pythagorean Theorem states: *If the square of the length of the longest side of a triangle is equal to the sum of the squares of the lengths of the other two sides, then the triangle is a right triangle.* Does the 3-4-5 method rely on the Pythagorean Theorem or on its converse? Explain.

3. How could you use the 3-4-5 method to mark out a rectangular garden area that measures 25 feet by 40 feet? (*Hint:* Begin by marking a figure that is 30 feet by 40 feet.)

Technology Activity Keystrokes

For use with page 542

TI-92

Construction

1. Draw △ABC.

[F3] 3 (Move cursor for location of A.) [ENTER] A (Move cursor for location of B.) [ENTER] (Move cursor for location of C.) [ENTER] C

2. Measure $\overline{AC}$, $\overline{BC}$, and $\overline{AB}$.

[F6] 1 (Place cursor on A.) [ENTER] (Move cursor to C.) [ENTER] (Move cursor to B.) [ENTER] (Move cursor to C.) [ENTER] (Move cursor to A.) [ENTER] (Move cursor to B.) [ENTER]

3. Calculate the values of $(AC)^2 + (BC)^2$ and $(AB)^2$.

[F6] 6 (Use cursor to highlight the length of AC.) [ENTER] [^] [2] [+] (Highlight the length of BC.) [ENTER] [^] [2] [ENTER] (Result will appear on the screen.) [F6] 6 (Use cursor to highlight the length of AB.) [ENTER] [^] [2] [ENTER]

4. Measure ∠C.

[F6] 3 (Place cursor on A.) [ENTER] (Move cursor to C.) [ENTER] (Move cursor to B.) [ENTER]

Investigate

1. Set up the table.

[APPS] 6 2

Cursor to Variable and choose sysdata from the menu list.

[F1] 8 [ENTER] (This will clear the contents of the data editor.)

[APPS] 8 1 (Return to the geometry window.)

Record the values of AC, BC, AB, $(AC)^2 + (BC)^2$, $(AB)^2$ and $m\angle C$.

[F6] 7 2 (Cursor to length AC.) [ENTER] (Cursor to length BC.) [ENTER] (Cursor to length AB.) [ENTER] (Cursor to the value of $(AC)^2 + (BC)^2$.) [ENTER] (Cursor to the value of $(AB)^2$.) [ENTER] (Cursor to $m\angle C$.) [ENTER]

[F6] 7 1 (Stores the values in sysdata.)

2. Drag point C.

[F1] 1 (Place cursor on point C.) [ENTER]

LESSON
9.3
CONTINUED

NAME _____ DATE _____

Technology Activity Keystrokes

For use with page 542

(Use the drag key 🖱 and the cursor pad to drag the point.)

Store the new values in sysdata.

F6 7 1

3. Repeat as many times as needed, making sure to store the new values when ∠C is an acute angle, a right angle, and an obtuse angle.

SKETCHPAD

Construct

1. Draw △ABC using the segment straightedge tool.

2. Measure $\overline{AC}$, $\overline{BC}$, and $\overline{AB}$. Use the selection arrow tool to select $\overline{AC}$, $\overline{BC}$, and $\overline{AB}$. Then choose **Length** from the **Measure** menu.

3. Calculate the values of $(AC)^2 + (BC)^2$ and $(AB)^2$. Choose **Calculate** from the **Measure** menu. Click the length of $\overline{AC}$, click **^**, click 2, click **+**, click the length of $\overline{BC}$, click **^**, click 2, and click OK. Choose **Calculate** from the **Measure** menu. Click the length of $\overline{AB}$, click **^**, click 2, and click OK.

4. Measure ∠C. Use the selection arrow tool to select points B, C, and A (in that order). Choose **Angle** from the **Measure** menu.

Investigate

1. Set up the table. Use the selection arrow tool to select the lengths of $\overline{AC}$, $\overline{BC}$, and $\overline{AB}$, the values of $(AC)^2 + (BC)^2$ and $(AB)^2$, and the measure of ∠C. Then choose **Tabulate** from the **Measure** menu.

2. Using the translate selection arrow, drag point C. To add the new values to the table, select the table and choose **Add Entry** from the **Measure** menu.

3. Repeat as many times as needed, making sure to store the new values when ∠C is an acute angle, a right angle, and an obtuse angle.

For use with page 547, Exercise 38.

TI-92

1. *Nonspecial quadrilateral* Use the polygon command (**F3** 4). [1]

2. *Parallelogram* Draw a segment (**F2** 5). Draw a point not on the segment (**F2** 1). Draw a line parallel to the segment through the point not on the segment (**F4** 2). Draw a segment from an endpoint of the first segment to a point on the parallel line (**F2** 5). Draw a line parallel to this new segment that goes through the other endpoint of the first segment (**F4** 2).[1]

Geometry
Chapter 9 Resource Book

LESSON

9.3

CONTINUED

NAME _____ DATE _____

Technology Activity Keystrokes

For use with page 547

3. *Rhombus* First draw $\overline{AB}$ (F2 5). Use the compass command (F4 8) to construct a circle with A as the center and $\overline{AB}$ as the radius. Draw a segment from A to a point on the circle and label the point C. Draw a line parallel to $\overline{AB}$ through point C (F4 2). Use the compass command (F4 8) to construct a circle with C as the center and $\overline{AC}$ as the radius. Label the intersection of the new circle and the parallel line point D (F2 3). Draw the rhombus using the polygon command (F3 4) to connect points A, B, C, and D. [1]

4. *Square* Use the regular polygon command [F3 5 (Locate a point in the center of the screen.) ENTER (Move the cursor to indicate the size of the figure.) ENTER (Move the cursor until the number in braces is a 4.) ENTER]. [1]

5. *Rectangle* First draw $\overline{AB}$ (F2 5). Draw perpendicular lines to $\overline{AB}$ through A and B (F4 1). Draw point C on one of the perpendicular lines (F2 2) and then draw a line parallel to $\overline{AB}$ through C (F4 2). [1]

[1]Now draw the diagonals of the figure, then measure the sides and diagonals using the distance and length command (F6 1).

SKETCHPAD

1. *Nonspecial quadrilateral* Use the segment straightedge tool. [2]

2. *Parallelogram* Draw $\overline{EF}$. Draw point G not on $\overline{EF}$. Draw a line parallel to $\overline{EF}$ through G. Draw a segment from E to a point on the parallel line. Draw a line parallel to this segment through F. [2]

3. *Rhombus* Draw a circle with center I and a point on the circle, J. Draw $\overline{IJ}$. Draw $\overline{IK}$ with K on the circle. Draw a line parallel to $\overline{IJ}$ through K. Draw a line parallel to $\overline{IK}$ through J. [2]

4. *Square* Draw a circle with center M and a point on the circle, N. Draw $\overline{MN}$. Draw lines perpendicular to $\overline{MN}$ through points M and N. Label the intersection of the perpendicular line and the circle point O. Draw a line parallel to $\overline{MN}$ through O. [2]

5. *Rectangle* Draw $\overline{QR}$. Draw lines perpendicular to $\overline{QR}$ through Q and R. Draw point S on one of the perpendicular lines. Draw a line parallel to $\overline{QR}$ through S. [2]

[2]Draw the diagonals of the figure. Select the sides and the diagonals, then choose **Length** from the **Measure** menu. If necessary, select the endpoints of the segment, then choose **Distance** from the **Measure** menu.

Lesson 9.3

Practice A
For use with pages 543–549

Decide whether the numbers can represent the side lengths of a triangle.

1. 5, 4, 3

2. 5, 6, 7

3. 5, 5, 10

4. 5, 10, 10

5. 5, 10, 15

6. 5, 15, 15

Tell whether the triangle is a right triangle.

7.

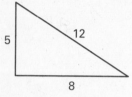

8.

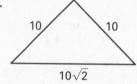

9.

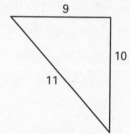

10.

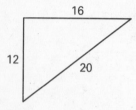

11.

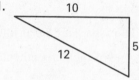

12.

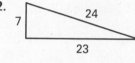

Classify the triangles with the given side lengths as *right*, *acute*, or *obtuse*.

13. 6, 8, 10

14. 6, 6, 10

15. 6, 10, 10

16. $\sqrt{6}$, $\sqrt{8}$, $\sqrt{10}$

17. 0.6, 0.8, 1.0

18. 7, 9, 11

Classify the quadrilateral. Explain how you can prove that the quadrilateral is that type.

19.

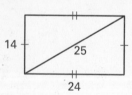

20.

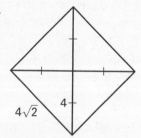

21.

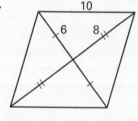

In Exercises 22–24, you will use two different methods for determining whether △*ABC* is a right triangle.

22. *Method 1* Find the slope of $\overline{AC}$ and the slope of $\overline{BC}$. What do the slopes tell you about ∠*ACB*? Is △*ABC* a right triangle? How do you know?

23. *Method 2* Use the Distance Formula and the Converse of the Pythagorean Theorem to determine whether △*ABC* is a right triangle.

24. Which method would you use to determine whether a given triangle is right, acute, or obtuse? Explain.

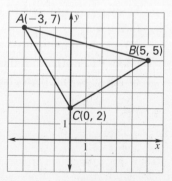

NAME _____ DATE _____

Practice B

For use with pages 543–549

Tell whether the triangle is a right triangle.

1.
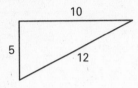

2.
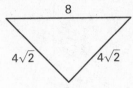

3.
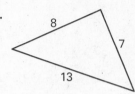

4.
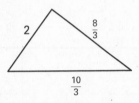

5.
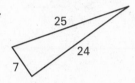

6.
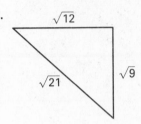

Decide whether the numbers can represent the side lengths of a triangle. If they can, classify the triangle as *right*, *acute*, or *obtuse*.

7. 5, 12, 13

8. $\sqrt{8}$, 4, 6

9. 20, 21, 28

10. 15, 36, 39

11. $\sqrt{13}$, 10, 12

12. 14, 48, 50

Classify the quadrilateral. Explain how you can prove that the quadrilateral is that type.

13.
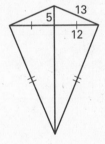

14.
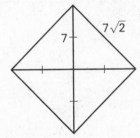

15.
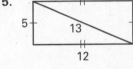

16. *Deck*   A contractor is building a deck adjacent to a home as shown. How can he be sure that the deck is square (the corners are right angles) when he lost his t-square and only has a tape measure? Explain your reasoning.

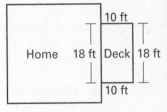

Roof   In Exercises 17 and 18, use the diagram and the following information.

The slope of the roof is $\dfrac{5}{12}$. The height of the roof is 15 feet.

17. What is the length from gutter to peak of the roof?

18. If a row of shingles is 5 inches high, how many rows of shingles are needed for one side of the roof?

Tell whether the triangle is a right triangle.

1.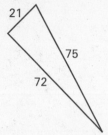
21
75
72

2.
$2\sqrt{17}$ $6\sqrt{2}$
2

3.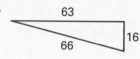
63
16
66

4.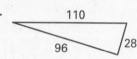
110
96 28

5.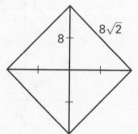
4.3
4.4
4.5

6.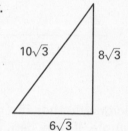
$10\sqrt{3}$ $8\sqrt{3}$
$6\sqrt{3}$

Decide whether the numbers can represent the side lengths of a triangle. If they can, classify the triangle as *right*, *acute*, or *obtuse*.

7. $7, \sqrt{5}, 3\sqrt{6}$

8. $8, 12, 18$

9. $6, \sqrt{15}, 5\sqrt{2}$

10. $7, 11, 20$

11. $5, 7, \sqrt{74}$

12. $21, 72, 75$

Classify the quadrilateral. Explain how you can prove that the quadrilateral is that type.

13.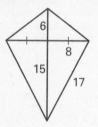
6
8
15
17

14.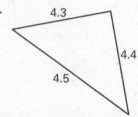
8 $8\sqrt{2}$

15.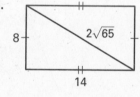
8 $2\sqrt{65}$
14

16. Quadrilateral *QUAD* has vertices at $Q = (-5, 2)$, $U = (-1, 7)$, $A = (4, 3)$, and $D = (0, -2)$. Plot the figure and indicate what type of quadrilateral *QUAD* is. Find the perimeter of *QUAD*.

Write a two-column proof or a paragraph proof.

17. **Given:** $AB = 6$, $BC = 8$, $AC = 10$

Prove: $\angle 1$ is a right angle

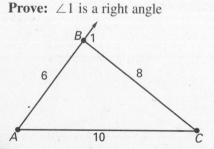
B 1
6 8
A 10 C

18. **Given:** $XZ = 3$, $ZY = 6$, $KY = 8$

Prove: $\angle 2$ is obtuse

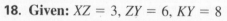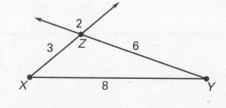
2
3 Z 6
X 8 Y

NAME _____ DATE _____

Reteaching with Practice

For use with pages 543–549

GOAL **Use the converse of the Pythagorean Theorem to solve problems and use side lengths to classify triangles by their angle measures**

> **Theorem 9.5   Converse of the Pythagorean Theorem**
> If the square of the length of the longest side of a triangle is equal to the sum of the squares of the lengths of the other two sides, then the triangle is a right triangle.
>
> **Theorem 9.6**
> If the square of the length of the longest side of a triangle is less than the sum of the squares of the lengths of the other two sides, then the triangle is acute.
>
> **Theorem 9.7**
> If the square of the length of the longest side of a triangle is greater than the sum of the squares of the lengths of the other two sides, then the triangle is obtuse.

EXAMPLE 1 *Verifying Right Triangles*

The triangles below appear to be right triangles. Tell whether they are right triangles.

a.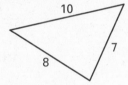

b.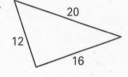

SOLUTION

Let c represent the length of the longest side of the triangle (you do not want to call this the "hypotenuse" because you do not yet know if the triangle is a right triangle). Check to see whether the side lengths satisfy the equation $c^2 = a^2 + b^2$.

a. $10^2 \overset{?}{=} 8^2 + 7^2$

    $100 \overset{?}{=} 64 + 49$

    $100 \neq 113$

The triangle is not a right triangle.

b. $20^2 \overset{?}{=} 12^2 + 16^2$

    $400 \overset{?}{=} 144 + 256$

    $400 = 400$

The triangle is a right triangle.

NAME _____ DATE _____

Reteaching with Practice

For use with pages 543–549

Exercises for Example 1

In Exercises 1–3, determine if the triangles are right triangles.

1.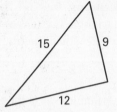

2.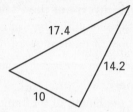

3.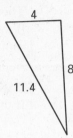

EXAMPLE 2 *Classifying Triangles*

Decide whether the set of numbers can represent the side lengths of a triangle. If they can, classify the triangle as *right*, *acute*, or *obtuse*.

a. 58, 69, 80   **b.** 11, 30, 39

SOLUTION

You can use the Triangle Inequality to confirm that each set of numbers can represent the side lengths of a triangle.

Compare the square of the length of the longest side with the sum of the squares of the lengths of the two shorter sides.

a. $c^2 \underline{\;?\;} a^2 + b^2$ Compare c^2 with $a^2 + b^2$.

   $80^2 \underline{\;?\;} 58^2 + 69^2$ Substitute.

   $6400 \underline{\;?\;} 3364 + 4761$ Multiply.

   $6400 < 8125$ c^2 is less than $a^2 + b^2$.

Because $c^2 < a^2 + b^2$, the triangle is acute.

b. $c^2 \underline{\;?\;} a^2 + b^2$ Compare c^2 with $a^2 + b^2$.

   $39^2 \underline{\;?\;} 11^2 + 30^2$ Substitute.

   $1521 \underline{\;?\;} 121 + 900$ Multiply.

   $1521 > 1021$ c^2 is greater than $a^2 + b^2$.

Because $c^2 > a^2 + b^2$, the triangle is obtuse.

Exercises for Example 2

Decide whether the set of numbers can represent the side lengths of a triangle. If they can, classify the triangle as *right*, *acute*, or *obtuse*.

4. 5, $\sqrt{56}$, 9   **5.** 23, 44, 70   **6.** 12, 80, 87   **7.** 4, 7, 10

Geometry
Chapter 9 Resource Book

NAME _____ DATE _____

Quick Catch-Up for Absent Students

For use with pages 542–549

The items checked below were covered in class on (date missed) _____

Activity 9.3: Investigating Sides and Angles of Triangles (p. 542)

_____ **Goal:** Use geometry software to explore how the angle measures of a triangle are related to its side lengths.

_____ Student Help: Software Help

Lesson 9.3: The Converse of the Pythagorean Theorem

_____ **Goal 1:** Use the Converse of the Pythagorean Theorem to solve problems. (p. 543)

Material Covered:

_____ Example 1: Verifying Right Triangles

_____ **Goal 2:** Use side lengths to classify triangles by their angle measures. (pp. 544–545)

Material Covered:

_____ Student Help: Look Back

_____ Example 2: Classifying Triangles

_____ Student Help: Look Back

_____ Example 3: Building a Foundation

_____ Other (specify) _____

_____

Homework and Additional Learning Support

_____ Textbook (specify) _pp. 545–549_____

_____

_____ *Reteaching with Practice* worksheet (specify exercises)_____

_____ *Personal Student Tutor* for Lesson 9.3

NAME _____ DATE _____

Interdisciplinary Application

For use with pages 543–549

Kitchen Space

HOME ECONOMICS   When cooking in your home, you need a certain amount of working space to make preparing a meal easy. Many people want a large kitchen in their home, but sometimes when a kitchen is too large, it makes the distances between the work surfaces and appliances too far apart to be convenient. Use a *work triangle* to help you decide what are the best distances between appliances. A work triangle in a kitchen is the triangle formed by the three lines that connect the center front of the refrigerator, sink, and stove.

The most desirable size of the work triangle is a total perimeter of 15 feet to 22 feet. The smallest acceptable perimeter is 12 feet and the largest is 26 feet. In addition, traffic through the kitchen should not cut through the work triangle area. Distances between appliances should be no more than 4 feet to 7 feet from refrigerator to sink, 4 feet to 6 feet from sink to stove, and 4 feet to 9 feet from stove to refrigerator.

In Exercises 1–6, use the information above and the diagrams below.

Kitchen 1

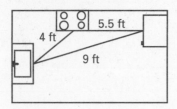

Kitchen 2

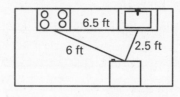

1. What is the perimeter of the work triangle in Kitchen 1? Is the work triangle within the desirable range?

2. What is the perimeter of the work triangle in Kitchen 2? Is the work triangle within the desirable range?

3. Which, if any, of the distances in the work triangles for Kitchen 1 and Kitchen 2 are outside the acceptable limits between appliances?

4. Show whether the work triangle in Kitchen 1 is a right triangle.

5. Show whether the work triangle in Kitchen 2 is a right triangle.

6. In Kitchen 1, you can slide the refrigerator down the wall 2.5 feet so that it is directly across from the sink. Does this improve the work triangle's perimeter and appliance distances? Is the new work triangle a right triangle?

Challenge: Skills and Applications

For use with pages 543–549

1. Here is a formula for generating Pythagorean triples. If m and n are positive integers, with $m < n$, let $a = n^2 - m^2$, $b = 2mn$, and $c = n^2 + m^2$.

   a. Show that a, b, and c form a Pythagorean triple.

   b. List the Pythagorean triples that are generated using $n \leq 5$.

   c. It can be shown that *every* Pythagorean triple can be generated in this manner. Find expressions for m and n in terms of a, b, and c.

   d. If you are given the three numbers of a Pythagorean triple and asked to find the corresponding values of m and n, how can you decide which number is a, which is b, and which is c?

   e. Find the values of m and n for the Pythagorean triple 56, 90, 106.

   f. Find the values of m and n for the Pythagorean triple 48, 55, 73.

2. Let *PQRS* be a parallelogram with side lengths $QR = PS = c$ and $QP = RS = d$, and diagonal lengths $PR = e$ and $QS = f$.

   a. Justify drawing auxiliary line segments $\overline{QT}$, $\overline{SU}$, and $\overline{UP}$, as shown.

   b. Use the Pythagorean Theorem and the properties of algebra to evaluate $e^2 + f^2$ in terms of c and d.

   c. Based on your work, write a general statement about the relationship between the lengths of the sides and the diagonals of a parallelogram.

   d. Using the diagram, show that the relationship you found in part (c) does *not* hold true for a kite.

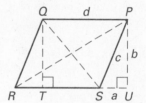

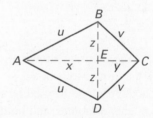

In Exercises 3–8, find the possible values of *x*.

3. $\triangle ABC$ is a right triangle; $AB = x$, $BC = x + 1$, $AC = x + 9$.

4. $\triangle DEF$ is a right triangle; $DE = 12$, $EF = x - 1$, $DF = x + 1$.

5. $\triangle GHI$ is a right triangle; $GH = 5$, $HI = x + 4$, $GI = 2x - 3$.

6. $\triangle JKL$ is a right triangle; $JK = 3x - 6$, $KL = 2x + 11$, $JL = 20$.

7. $\triangle MNO$ is an acute triangle; $MN = x - 1$, $NO = x + 1$, $MO = 8$.

8. $\triangle PQR$ is an obtuse triangle; $PQ = x$, $QR = x + 1$, $PR = 5$.

Quiz 1

For use after Lessons 9.1–9.3

In Exercises 1–4, use the diagram. *(Lesson 9.1)*

1. Write a similarity statement about the three triangles shown in the diagram.

2. Which segment's length is the geometric mean between *RS* and *ST*?

3. Find *TS*.

4. Find *SQ*.

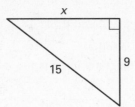

Answers

1. _____
2. _____
3. _____
4. _____
5. _____
6. _____
7. _____
8. _____

Find the unknown side length. Simplify answers that are radicals. *(Lesson 9.2)*

5.

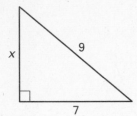

6.

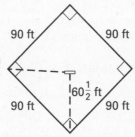

7.

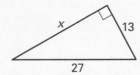

8. *Baseball Field*   The diagram shown at the right shows the dimensions of a baseball field. Is the triangle formed by third base, the pitcher's mound, and home plate a right triangle? Explain. *(Lesson 9.3)*

TEACHER'S NAME _____ CLASS _____ ROOM _____ DATE _____

Lesson Plan

2-day lesson (See *Pacing the Chapter,* TE pages 524C–524D)   **For use with pages 550–557**

  GOALS   1. **Find the side lengths of special right triangles.**
2. **Use special right triangles to solve real-life problems.**

State/Local Objectives _____

_____

✓ Check the items you wish to use for this lesson.

STARTING OPTIONS

_____ Homework Check: TE page 546: Answer Transparencies
_____ Warm-Up or Daily Homework Quiz: TE pages 551 and 549, CRB page 55, or Transparencies

TEACHING OPTIONS

_____ Motivating the Lesson: TE page 552
_____ Concept Activity: SE page 550
_____ Lesson Opener (Visual Approach): CRB page 56 or Transparencies
_____ Technology Activity with Keystrokes: CRB pages 41–43
_____ Examples: Day 1: 1–5: SE pages 551–553; Day 2: See the Extra Examples.
_____ Extra Examples: Day 1 or Day 2: 1–5, TE pages 552–553 or Transp.; Internet
_____ Closure Question: TE page 553
_____ Guided Practice: SE page 554   Day 1: Exs. 1–11; Day 2: See Checkpoint Exs. TE pages 552–553

APPLY/HOMEWORK

Homework Assignment

_____ Basic   Day 1: 12–26; Day 2: 27–38, 42–49
_____ Average   Day 1: 12–26; Day 2: 27–38, 42–49
_____ Advanced   Day 1: 12–26; Day 2: 27–49

Reteaching the Lesson

_____ Practice Masters: CRB pages 57–59 (Level A, Level B, Level C)
_____ Reteaching with Practice: CRB pages 60–61 or Practice Workbook with Examples
_____ Personal Student Tutor

Extending the Lesson

_____ Applications (Real-Life): CRB page 63
_____ Math & History: SE page 557; CRB page 64; Internet
_____ Challenge: SE page 556; CRB page 65 or Internet

ASSESSMENT OPTIONS

_____ Checkpoint Exercises: Day 1 or Day 2: TE pages 552–553 or Transp.
_____ Daily Homework Quiz (9.4): TE page 557, CRB page 68, or Transparencies
_____ Standardized Test Practice: SE page 556; TE page 557; STP Workbook; Transparencies

Notes _____

_____

_____

TEACHER'S NAME _____ CLASS _____ ROOM _____ DATE _____

Lesson Plan for Block Scheduling

1-day lesson (See *Pacing the Chapter*, TE pages 524C–524D)   **For use with pages 550–557**

GOALS
1. **Find the side lengths of special right triangles.**
2. **Use special right triangles to solve real-life problems.**

State/Local Objectives _____

_____

_____

✓ **Check the items you wish to use for this lesson.**

STARTING OPTIONS

____ Homework Check: TE page 546: Answer Transparencies
____ Warm-Up or Daily Homework Quiz: TE pages 551 and
        549, CRB page 55, or Transparencies

CHAPTER PACING GUIDE	
Day	**Lesson**
. 1	Assess Ch. 8; 9.1 (all)
2	9.2 (all); 9.3 (begin)
3	9.3 (end); **9.4 (begin)**
4	**9.4 (end)**; 9.5 (begin)
5	9.5 (end); 9.6 (all)
6	9.7 (all)
7	Review Ch. 9; Assess Ch. 9

TEACHING OPTIONS

____ Motivating the Lesson: TE page 552
____ Concept Activity: SE page 550
____ Lesson Opener (Visual Approach): CRB page 56 or Transparencies
____ Technology Activity with Keystrokes: CRB pages 41–43
____ Examples: Day 3: 1–5: SE pages 551–553; Day 4: See the Extra Examples.
____ Extra Examples: Day 3 or Day 4: 1–5, TE pages 552–553 or Transp.; Internet
____ Closure Question: TE page 553
____ Guided Practice: SE page 554   Day 3: Exs. 1–11; Day 4: See Checkpoint Exs. TE pages 552–553

APPLY/HOMEWORK

Homework Assignment (See also the assignments for Lessons 9.3 and 9.5.)

____ Block Schedule: Day 3: 12–26; Day 4: 27–38, 42–49

Reteaching the Lesson

____ Practice Masters: CRB pages 57–59 (Level A, Level B, Level C)
____ Reteaching with Practice: CRB pages 60–61 or Practice Workbook with Examples
____ Personal Student Tutor

Extending the Lesson

____ Applications (Real-Life): CRB page 63
____ Math & History: SE page 557; CRB page 64; Internet
____ Challenge: SE page 556; CRB page 65 or Internet

ASSESSMENT OPTIONS

____ Checkpoint Exercises: Day 3 or Day 4: TE pages 552–553 or Transp.
____ Daily Homework Quiz (9.4): TE page 557, CRB page 68, or Transparencies
____ Standardized Test Practice: SE page 556; TE page 557; STP Workbook; Transparencies

Notes _____

_____

_____

NAME _____ DATE _____

WARM-UP EXERCISES
For use before Lesson 9.4, pages 550–557

Solve the equation for the missing variable. Assume all variables are positive. Express the answer in simplified radical form.

1. $c^2 = 6^2 + 6^2$

2. $c^2 - 4^2 = 4^2$

3. $c^2 - 100 = (10\sqrt{3})^2$

4. $a^2 + 8^2 = 256$

5. $(18\sqrt{3})^2 + b^2 = 1296$

DAILY HOMEWORK QUIZ
For use after Lesson 9.3, pages 542–549

Decide whether the numbers can represent the side lengths of a triangle. If they can, classify the triangle as *right*, *acute*, or *obtuse*.

1. 12, 35, 37

2. 22, 25, 40

3. 15, 17, 34

Graph points *P*, *Q*, and *R*. Connect the points to form △*PQR*. Decide whether △*PQR* is *right*, *acute*, or *obtuse*.

4. $P(-3, 2), Q(-2, -2), R(1, 1)$

5. $P(1, 2), Q(5, -2), R(3, -4)$

Use the square pattern shown. Ignore the dots until Exercise 3.

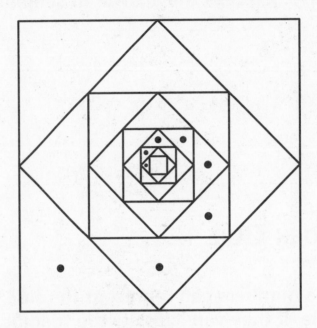

1. Identify 9 nested squares in the pattern. How does the area of one square compare to the area of the next smaller square?

2. Identify 32 similar triangles in the pattern. How many different sizes are there? How many are there of each size? What are the angle measures of each triangle? How does a leg of one triangle compare to the hypotenuse of the next smaller triangle?

3. Color the pattern so that four spirals result. (*Hint:* Use one color for all the triangles that contain a dot. This is one spiral.)

4. Create this pattern on graph paper. First draw a square with sides 32 units long. Find the midpoints of the sides and connect them to draw the next square. Continue until you reach the center. Color your pattern in any way that is different from Exercise 3.

NAME _____ DATE _____

Practice A

For use with pages 551–557

Find the value of each variable in the polygon.

1. Equilateral △ABC

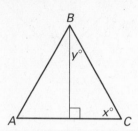

2. Square ABCD

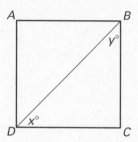

3. Regular hexagon ABCDEF

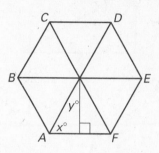

Find the value of each variable. Write answers in simplest radical form.

4.

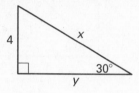

5.

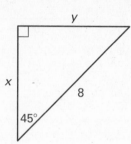

6.

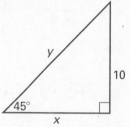

7.

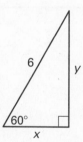

8.

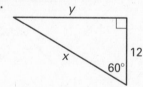

9.

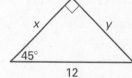

Sketch the figure that is described. Find the requested length. Round decimals to the nearest tenth.

10. The side length of an equilateral triangle is 20 centimeters. Find the length of an altitude of the triangle.

11. The perimeter of a square is 20 centimeters. Find the length of a diagonal.

12. The diagonal of a square is 10 inches. Find the length of a side.

Baseball **In Exercises 13–15, use the diagram and the following information.**

The infield of a baseball field is a square. The distance from home plate to first base is 90 feet.

13. What is the distance from home plate to second base?

14. What is the distance from third base to first base?

15. If the pitcher's mound is 60 feet 6 inches from home plate, is it the midpoint of the diagonal from home plate to second base?

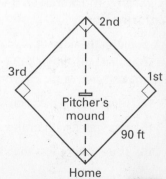

Find the value of each variable. Write answers in simplest radical form.

1.
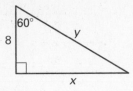

2.
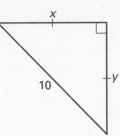

3.
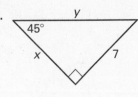

4.
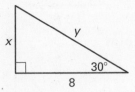

5.
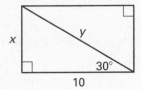

6.
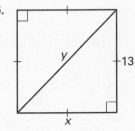

7.
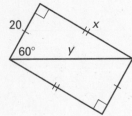

8.
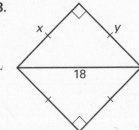

9.
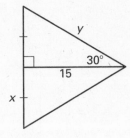

Sketch the figure that is described. Find the requested length. Round decimals to the nearest tenth.

10. The perimeter of a square is 20 centimeters. Find the length of a diagonal.

11. The altitude of an equilateral triangle is 18 inches. Find the length of a side.

12. The hypotenuse of an isosceles right triangle is 16 centimeters. Find the length of a side.

13. The length of the diagonal of a square is $\dfrac{5\sqrt{2}}{2}$. Find the length of a side.

Canyon **In Exercises 14–16, use the diagram and the following information.**

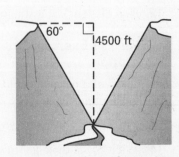

A point on the edge of a symmetrical canyon is 4500 feet above a river that cuts through the canyon floor. The angle of depression from each side of the canyon to the canyon floor is 60°.

14. Find the distance across the canyon.

15. Find the length of the canyon wall (from the edge to the river).

16. Is it more or less than a mile across the canyon? (5280 feet = 1 mile)

Lesson 9.4

NAME _____ DATE _____

Practice C

For use with pages 551–557

Find the value of each variable. Write answers in simplest radical form.

1.
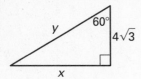

2.
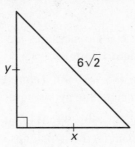

3.
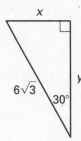

4.
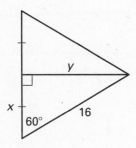

5.
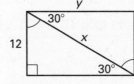

6.
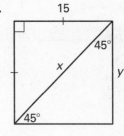

7.
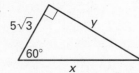

8.
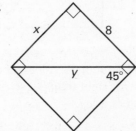

9.
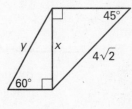

Sketch the figure that is described. Find the requested length, perimeter, or area. Round decimals to the nearest tenth.

10. The altitude of an equilateral triangle is 12 centimeters. Find the perimeter of the triangle.

11. The diagonal of a square is 8 inches. Find the area.

12. The perimeter of a rectangle is 66 centimeters. The length is twice the width. Find the length of the diagonal.

13. The perimeter of an equilateral triangle is 36 inches. Find the length of an altitude.

14. Each figure below is a 30°-60°-90° triangle. Find the value of x.

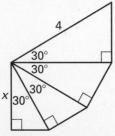

15. Each figure below is a 45°-45°-90° triangle. Find the value of x.

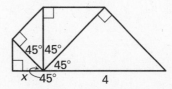

NAME _____ DATE _____

Reteaching with Practice

For use with pages 551–557

GOAL Find the side lengths of special right triangles

VOCABULARY

Right triangles whose angle measures are 45°- 45°- 90° or 30°- 60°- 90° are called **special right triangles.**

Theorem 9.8   The 45°- 45°- 90° Triangle Theorem
In a 45°- 45°- 90° triangle, the hypotenuse is $\sqrt{2}$ times as long as each leg.

Theorem 9.9   The 30°- 60°- 90° Triangle Theorem
In a 30°- 60°- 90° triangle, the hypotenuse is twice as long as the shorter leg, and the longer leg is $\sqrt{3}$ times as long as the shorter leg.

EXAMPLE 1   *Finding Side Lengths in a 45°- 45°- 90° Triangle*

Find the value of x.

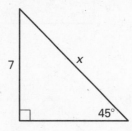

SOLUTION

By the Triangle Sum Theorem, the measure of the third angle is 45°. The triangle is a 45°- 45°- 90° right triangle, so the length x of the hypotenuse is $\sqrt{2}$ times the length of a leg.

$$\text{Hypotenuse} = \sqrt{2} \cdot \text{leg} \qquad \text{45°- 45°- 90° Triangle Theorem}$$
$$x = \sqrt{2} \cdot 7 \qquad \text{Substitute.}$$
$$x = 7\sqrt{2} \qquad \text{Simplify.}$$

Exercises for Example 1

Find the value of each variable.

1.
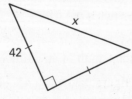

2.
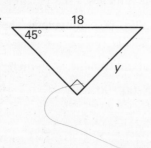

3.
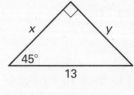

Reteaching with Practice

For use with pages 551–557

EXAMPLE 2 *Finding Side Lengths in a 30°-60°-90° Triangle*

Find the value of *x*.

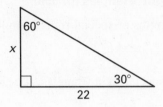

SOLUTION

Because the triangle is a 30°- 60°- 90° triangle, the longer leg is $\sqrt{3}$ times the length *x* of the shorter leg.

Longer leg $= \sqrt{3} \cdot$ shorter leg	30°- 60°- 90° Triangle Theorem
$22 = \sqrt{3} \cdot x$	Substitute.
$\dfrac{22}{\sqrt{3}} = x$	Divide each side by $\sqrt{3}$.
$\dfrac{\sqrt{3}}{\sqrt{3}} \cdot \dfrac{22}{\sqrt{3}} = x$	Multiply numerator and denominator by $\sqrt{3}$.
$\dfrac{22\sqrt{3}}{3} = x$	Simplify.

Exercises for Example 2

Find the value of each variable.

4.

5.

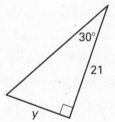

6.

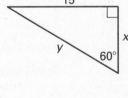

NAME _____ DATE _____

Quick Catch-Up for Absent Students

For use with pages 550–557

The items checked below were covered in class on (date missed) _____

Activity 9.4: Investigating Special Right Triangles (p. 550)

_____ **Goal:** Determine how the side lengths of special right triangles are related.

Lesson 9.4: Special Right Triangles

_____ **Goal 1:** Find the side lengths of special right triangles. (pp. 551–552)

Material Covered:

_____ Example 1: Finding the Hypotenuse in a 45°- 45°- 90° Triangle

_____ Example 2: Finding a Leg in a 45°- 45°- 90° Triangle

_____ Example 3: Side Lengths in a 30°- 60°- 90° Triangle

Vocabulary:

special right triangles, p. 551

_____ **Goal 2:** Use special right triangles to solve real-life problems. (p. 553)

Material Covered:

_____ Example 4: Finding the Height of a Ramp

_____ Example 5: Finding the Area of a Sign

_____ Other (specify) _____

_____

Homework and Additional Learning Support

_____ Textbook (specify) _pp. 554–557_____

_____

_____ Internet: Extra Examples at www.mcdougallittell.com

_____ *Reteaching with Practice* worksheet (specify exercises)_____

_____ *Personal Student Tutor* for Lesson 9.4

NAME _____ DATE _____

Real-Life Application:
When Will I Ever Use This?

For use with pages 551–557

Modular Homes

For many families, affordable housing is difficult to find. One option is a modular home. A modular home is built in pieces in a factory, transported to the new address, and assembled there. Once it arrives, it requires finish work such as attaching moldings, sealing seams, and final painting. It is more afford-able than a home built by contractors at the home site and can come in larger sizes than mobile homes since mobile homes are limited by what can safely be towed on the highway.

Roof pitch is a measure of the slope of the roof. Modular homes have a standard roof pitch of 6/12 or 12/12. The first number is the vertical change and the sec-ond number is the horizontal change. For example, a 6/12 pitch is a roof that has 6 feet of vertical rise for every 12 feet of horizontal distance.

In Exercises 1–6, use the following information.

You are purchasing a new modular home and want to be sure that there is plenty of attic space for storage and enough space to put an office in the attic at a later date. You can choose from the standard roof pitch available or other pitches.

1. Draw a right triangle that represents the 12/12 pitch.

2. What is the angle of pitch of the 12/12 roof?

3. Consider a pitch angle of 30°. With a base of 12, what is the vertical rise?

4. Consider a pitch angle of 60°. With a base of 12, what is the vertical rise?

5. From Exercises 2–4, which pitch angle gives the most headroom in the attic?

6. You are concerned about being able to safely climb onto the roof in the event that repairs are needed. A roof with an 8/12 pitch is about the steepest roof that can be safely climbed without roofing and safety equipment. After considering the amount of headroom in the attic and the safety in climbing the roof, what roof pitch should you choose?

NAME _____ DATE _____

Math and History Application

For use with page 557

HISTORY   James A. Garfield (1831–1881), the twentieth President of the United States, was not the typical president. While United States presidents are rarely known for their mathematical abilities, Garfield was an exception.

James Garfield was educated at Western Reserve Academy and Hiram College in Ohio. He graduated from Williams College in Massachusetts in 1856. He returned to Hiram to teach mathematics and became the principal of the school in 1858. In 1859, he left the school to become a member of the Ohio Senate. When the Civil War erupted in 1861, Garfield joined the Union Army. After serving in the Army, Garfield joined the House of Representatives where he focused on reforming the South. In 1880, James Garfield was elected President, but only four months later was shot and later died of complications. In his inaugural address, Garfield stressed the importance of education and the need to provide education for all Americans.

MATH   Although Garfield did not have a long presidency, he did leave his mark in mathematics. He developed his own proof of the Pythagorean Theorem. Garfield's proof of the Pythagorean Theorem was published in the New England Journal of Education in 1876. Garfield used the construction below to prove that $a^2 + b^2 = c^2$. The key to Garfield's proof is the area of a trapezoid, which is half the sum of the bases times the altitude.

In Exercises 1–3, use the diagram at the right. The diagram is based on the one drawn by James Garfield in his proof of the Pythagorean Theorem.

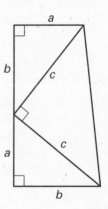

1. Write an expression for the area of the trapezoid in terms of a and b using the formula for the area of a trapezoid.

2. Write an expression in terms of a, b, and c for the combined areas of the three triangles.

3. Use the expressions in Exercises 1 and 2 to prove the Pythagorean Theorem.

NAME _____ DATE _____

Challenge: Skills and Applications

For use with pages 551–557

1. Refer to the diagram. Find the exact values of *p*, *q*, *r*, *s*, *t*, *u*, *v*, and *w*.

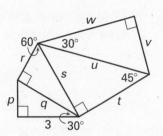
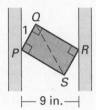

2. A brick is wedged between two parallel wooden planks that are 9 inches apart, as shown. If $m\angle 1 = m\angle RQS = 30°$, what is the length *QR* of the brick?

In Exercises 3–4, find the height *h* of the trapezoid in terms of the base lengths *a* and *b*. Rationalize the denominator.

3.

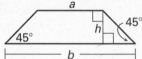

4.

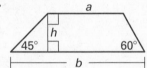

5. Let *YZ* = 2. Complete the following steps to find the side lengths of △*VWX*, a 15°-75°-90° triangle.

   a. Find *XY*, *VY*, and *VZ*.

   b. What kind of special right triangle is △*VWZ*? Find the lengths of the sides of △*VWZ*.

   c. Find the lengths of the sides of △*VWX*.

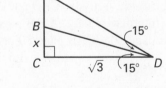

6. Here is another approach to finding the side lengths of a 15°-75°-90° triangle. Let $CD = \sqrt{3}$.

   a. Find *AB* and *AD* (in terms of *x*, where necessary).

   b. Write and solve a proportion to find the value of *x*. (*Hint:* Use a theorem in Lesson 8.6.)

   c. What are the lengths of the sides of △*BCD*?

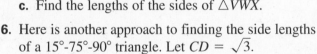

7. Using the side lengths you found in Exercises 5 and 6, use a calculator to verify numerically that △*BCD* ~ △*VWX*.

TEACHER'S NAME _____ CLASS _____ ROOM _____ DATE _____

Lesson Plan

2-day lesson (See *Pacing the Chapter,* TE pages 524C–524D)   For use with pages 558–566

GOALS 1. **Find the sine, the cosine, and the tangent of an acute angle.**
          2. **Use trigonometric ratios to solve real-life problems.**

State/Local Objectives _____

_____

✓ **Check the items you wish to use for this lesson.**

STARTING OPTIONS
____ Homework Check: TE page 554: Answer Transparencies
____ Warm-Up or Daily Homework Quiz: TE pages 558 and 557, CRB page 68, or Transparencies

TEACHING OPTIONS
____ Lesson Opener (Activity): CRB page 69 or Transparencies
____ Technology Activity with Keystrokes: CRB pages 70–71
____ Examples: Day 1: 1–7: SE pages 558–561; Day 2: See the Extra Examples.
____ Extra Examples: Day 1 or Day 2: 1–7, TE pages 559–561 or Transp.; Internet
____ Closure Question: TE page 561
____ Guided Practice: SE page 562   Day 1: Exs. 1–9; Day 2: See Checkpoint Exs. TE pages 559–561

APPLY/HOMEWORK
Homework Assignment
____ Basic   Day 1: 10–38 even; Day 2: 11–39 odd, 47–51, 53, 54, 56–60; Quiz 2: 1–7
____ Average   Day 1: 10–38 even, 40, 42, 46; Day 2: 11–39 odd, 43, 44, 47–51, 53, 54, 56–60;
           Quiz 2: 1–7
____ Advanced   Day 1: 10–38 even, 40–42, 46; Day 2: 11–39 odd, 43, 44, 47–60; Quiz 2: 1–7

Reteaching the Lesson
____ Practice Masters: CRB pages 72–74 (Level A, Level B, Level C)
____ Reteaching with Practice: CRB pages 75–76 or Practice Workbook with Examples
____ Personal Student Tutor

Extending the Lesson
____ Cooperative Learning Activity: CRB page 78
____ Applications (Interdisciplinary): CRB page 79
____ Challenge: SE page 565; CRB page 80 or Internet

ASSESSMENT OPTIONS
____ Checkpoint Exercises: Day 1 or Day 2: TE pages 559–561 or Transp.
____ Daily Homework Quiz (9.5): TE page 566, CRB page 84, or Transparencies
____ Standardized Test Practice: SE page 565; TE page 566; STP Workbook; Transparencies
____ Quiz (9.4–9.5): SE page 566; CRB page 81

Notes _____

_____

_____

TEACHER'S NAME _____ CLASS _____ ROOM _____ DATE _____

Lesson Plan for Block Scheduling

1-day lesson (See *Pacing the Chapter*, TE pages 524C–524D)   **For use with pages 558–566**

 GOALS
1. **Find the sine, the cosine, and the tangent of an acute angle.**
2. **Use trigonometric ratios to solve real-life problems.**

State/Local Objectives _____

_____

_____

✓ **Check the items you wish to use for this lesson.**

CHAPTER PACING GUIDE	
Day	**Lesson**
1	Assess Ch. 8; 9.1 (all)
2	9.2 (all); 9.3 (begin)
3	9.3 (end); 9.4 (begin)
4	9.4 (end); **9.5 (begin)**
5	**9.5 (end)**; 9.6 (all)
6	9.7 (all)
7	Review Ch. 9; Assess Ch. 9

STARTING OPTIONS
____ Homework Check: TE page 554: Answer Transparencies
____ Warm-Up or Daily Homework Quiz: TE pages 558 and
       557, CRB page 68, or Transparencies

TEACHING OPTIONS
____ Lesson Opener (Activity): CRB page 69 or Transparencies
____ Technology Activity with Keystrokes: CRB pages 70–71
____ Examples: Day 4: 1–7: SE pages 558–561; Day 5: See the Extra Examples.
____ Extra Examples: Day 4 or Day 5: 1–7, TE pages 559–561 or Transp.; Internet
____ Closure Question: TE page 561
____ Guided Practice: SE page 562   Day 4: Exs. 1–9; Day 5: See Checkpoint Exs. TE pages 559–561

APPLY/HOMEWORK
Homework Assignment (See also the assignments for Lessons 9.4 and 9.6.)
____ Block Schedule: Day 4: 10–38 even; Day 5: 11–39 odd, 43, 44, 47–51, 53, 54, 56–60; Quiz 2: 1–7

Reteaching the Lesson
____ Practice Masters: CRB pages 72–74 (Level A, Level B, Level C)
____ Reteaching with Practice: CRB pages 75–76 or Practice Workbook with Examples
____ Personal Student Tutor

Extending the Lesson
____ Cooperative Learning Activity: CRB page 78
____ Applications (Interdisciplinary): CRB page 79
____ Challenge: SE page 565; CRB page 80 or Internet

ASSESSMENT OPTIONS
____ Checkpoint Exercises: Day 4 or Day 5: TE pages 559–561 or Transp.
____ Daily Homework Quiz (9.5): TE page 566, CRB page 84, or Transparencies
____ Standardized Test Practice: SE page 565; TE page 566; STP Workbook; Transparencies
____ Quiz (9.4–9.5): SE page 566; CRB page 81

Notes _____

_____

_____

NAME _____ DATE _____

WARM-UP EXERCISES

For use before Lesson 9.5, pages 558–566

Solve each equation.

1. $0.875 = \dfrac{x}{18}$ **2.** $\dfrac{24}{y} = 0.5$ **3.** $\dfrac{y}{25} = 0.96$

4. $0.866x = 12$ **5.** $0.5x = 18$

DAILY HOMEWORK QUIZ

For use after Lesson 9.4, pages 550–557

**Find the value of each variable. Write answers in
simplest form.**

1.

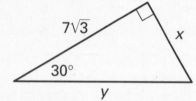

**Sketch the figure that is described. Find the requested
measure. Round decimals to the nearest tenth.**

2. The perimeter of a square is 48 meters. Find the length
of a diagonal.

3. An equilateral triangle has a side length of 10 inches.
Find the area of the triangle.

NAME _____ DATE _____

Activity Lesson Opener

For use with pages 558–566

SET UP: Work in a group.
You will need: • ruler • protractor • calculator

1. Each member of the group draws and labels a triangle as shown at the right. Choose a measure for $\angle A$, between 20° and 70°, which everyone in the group uses. Measure the sides of your triangle to the nearest millimeter. Are the triangles in your group congruent? Are they similar?

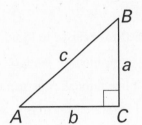

2. As a group, complete the table below by using measurements from your triangles. (Use a calculator to find the ratios, rounding to the nearest thousandth.)

Student's Name	a	b	c	$\dfrac{a}{c}$	$\dfrac{b}{c}$	$\dfrac{a}{b}$

3. Analyze your table. What do you notice about the column of values for each ratio? Why should the values be equal? Does the value of a ratio depend on the particular triangle measured?

4. Compare your results with other groups in the class. Does the value of a ratio depend on the measure of $\angle A$? Explain.

NAME _____ DATE _____

Technology Activity Keystrokes

For use with page 564

Keystrokes for Exercise 45
TI-92

1. Draw $\overline{AB}$.

 [F2] 5 (Move cursor to location for point *A*.) [ENTER] *A* (Move cursor to location for point *B*.) [ENTER] *B*

2. Draw a line through *B* perpendicular to $\overline{AB}$.

 [F4] 1 (Move cursor to *B*.) [ENTER] [ENTER]

3. Construct △*ABC* by drawing $\overline{AC}$ where *C* is above *B* and lies on the line perpendicular to $\overline{AB}$.

 [F2] 6 (Move cursor to *A*.) [ENTER] (Move cursor to line perpendicular to $\overline{AB}$ and above *B*.) [ENTER] *C*

4. Measure ∠*C* and ∠*A*.

 [F6] 3 (Move cursor to *A*.) [ENTER] (Move cursor to *C*.) [ENTER] (Move cursor to *B*.) [ENTER] [F6] 3 (Move cursor to *B*.) [ENTER] (Move cursor to *A*.) [ENTER] (Move cursor to *C*.) [ENTER]

5. Calculate the tangent of ∠*C* and ∠*A*. [F6] 6 [TAN] (Cursor to measure of ∠*C*.) [ENTER] [ )] [ENTER] (The result will appear on the screen.) [TAN] (Cursor to measure of ∠*C*.) [ENTER] [ )] [ENTER] (The result will appear on the screen.)

6. Drag ∠*C* and notice the results.

 [F1] 1 (Move cursor to *C* until prompt says "THIS POINT".) [ENTER] (Use the drag key 🖐 and the cursor pad to drag the point.)

Technology Activity Keystrokes

For use with page 564

SKETCHPAD

1. Draw $\overline{AB}$ using the segment straightedge tool.

2. Draw a line through B perpendicular to $\overline{AB}$. Choose the selection arrow tool, select point B, hold down the shift key and select $\overline{AB}$, and choose **Perpendicular Line** from the **Construct** menu.

3. Construct $\triangle ABC$ by drawing $\overline{AC}$ where C is above B and lies on the line perpendicular to $\overline{AB}$. Choose the segment straightedge tool.

4. Measure $\angle C$ and $\angle A$.

   For $\angle C$, choose the selection arrow tool, select point A, hold down the shift key and select point C and B (in that order).

   Choose **Angle** from the **Measure** menu. For $\angle A$, select point B, hold down the shift key and select points A and C (in that order). Choose **Angle** from the **Measure** menu.

5. Calculate the tangent of $\angle C$ and $\angle A$. Choose **Calculate . . .** from the **Measure** menu. In the Calculate dialog box, choose **tan** (from the **Functions** drop down menu). Then click the measure of $\angle C$, click ⬛ **)** , and click OK.

   Repeat this process for $\angle A$.

6. Drag $\angle C$ using the translate selection arrow tool and notice the results.

NAME _____ DATE _____

Practice A

For use with pages 558–566

Use the diagrams at the right to find the trigonometric ratio.

1. sin A
2. cos A
3. tan B
4. sin J
5. cos K
6. tan K

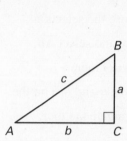

Find the sine, the cosine, and the tangent of the acute angles of the triangle. Express each value as a decimal rounded to four places.

7.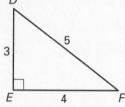

8.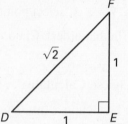

9.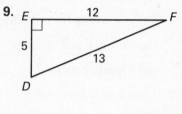

Use a calculator to approximate the given value to four decimal places.

10. sin 30°

11. cos 18°

12. tan 72°

13. sin 48°

14. tan 42°

15. cos 65°

16. tan 14°

17. sin 83°

Fill in the blank. Solve for the variable. Round decimals to the nearest tenth.

18. $\sin 52° = \dfrac{x}{?}$

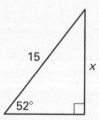

19. $\cos \underline{\ ?\ }° = \dfrac{x}{14}$

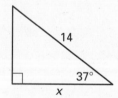

20. $\tan 24° = \dfrac{8}{?}$

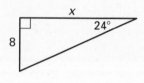

Find the value of each variable. Round decimals to the nearest tenth.

21.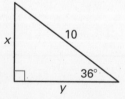

22.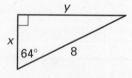

23.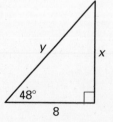

NAME _____ DATE _____

Practice B
For use with pages 558–566

Find the sine, the cosine, and the tangent of the acute angles of the triangle. Express each answer as a decimal rounded to four places.

1.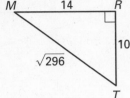

2.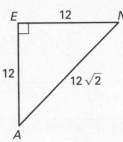

3.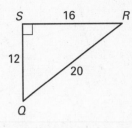

4.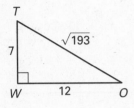

5.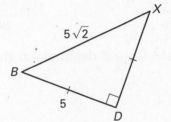

6.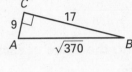

Use a calculator to approximate the given value to four decimal places.

7. $\sin 10°$

8. $\cos 38°$

9. $\tan 44°$

10. $\sin 74°$

11. $\tan 65°$

12. $\cos 63°$

13. $\sin 57°$

14. $\cos 33°$

Find the value of each variable. Round decimals to the nearest tenth.

15.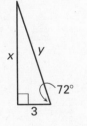

16.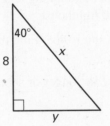

17.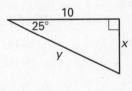

18.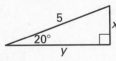

19.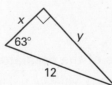

20.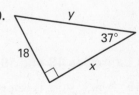

21. **Train**   A train is traveling up a slight grade with an angle of inclination of only 2°. After traveling 1 mile what is the vertical change in feet?

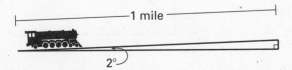

LESSON 9.5

Practice C

For use with pages 558–566

Find the sine, the cosine, and the tangent of the acute angles of the triangle. Express each answer as a decimal rounded to four places.

1.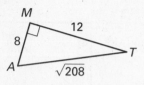

2.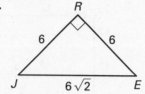

3.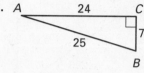

Use a calculator to approximate the given value to four decimal places.

4. sin 49°

5. cos 83°

6. tan 4°

7. sin 71°

8. tan 75°

9. cos 15°

10. sin 32°

11. cos 64°

Find the value of each variable. Round decimals to the nearest tenth.

12.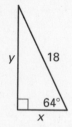

13.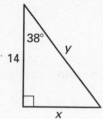

14.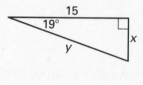

In Exercises 15–17, use the figure of the lighthouse.

15. At 2 P.M. the shadow of a lighthouse is 22 feet long and the angle of elevation is 72°. Find the height of the lighthouse.

16. At 4 P.M. the angle of elevation of the sun is 40°. Find the length of the shadow cast by the lighthouse.

17. At 6 P.M. will the length of the shadow be longer or shorter than it was at 4 P.M.? Explain.

In Exercises 18 and 19, use the figure of the escalator.

18. A new store is being built. An escalator is planned. It will make an angle of 34° with the floor. If the vertical distance between floors is 14 feet, how long will the escalator be?

19. If the angle made with the floor is changed to 36°, will the length of the escalator increase or decrease? Explain.

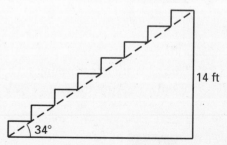

NAME _____ DATE _____

Reteaching with Practice

For use with pages 558–565

GOAL Find the sine, the cosine, and the tangent of an acute angle and use trigonometric ratios to solve real-life problems

VOCABULARY

A **trigonometric ratio** is a ratio of the lengths of two sides of a right triangle. The three basic trigonometric ratios are **sine, cosine,** and **tangent,** which are abbreviated as *sin, cos,* and *tan,* respectively.

The angle that your line of sight makes with a line drawn horizontally is called the **angle of elevation.**

Trigonometric Ratios

Let $\triangle ABC$ be a right triangle. The sine, the cosine, and the tangent of the acute angle $\angle A$ are defined as follows.

$$\sin A = \frac{\text{side opposite } \angle A}{\text{hypotenuse}} = \frac{a}{c}$$

$$\cos A = \frac{\text{side adjacent } \angle A}{\text{hypotenuse}} = \frac{b}{c}$$

$$\tan A = \frac{\text{side opposite } \angle A}{\text{side adjacent } \angle A} = \frac{a}{b}$$

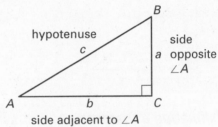

EXAMPLE 1 *Finding Trigonometric Ratios*

Find the sine, the cosine, and the tangent of the indicated angle.

a. $\angle A$

b. $\angle B$

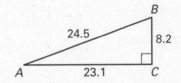

SOLUTION

a. The length of the hypotenuse is 24.5. For $\angle A$, the length of the opposite side is 8.2, and the length of the adjacent side is 23.1.

$$\sin A = \frac{\text{opp.}}{\text{hyp.}} = \frac{8.2}{24.5} \approx 0.3347$$

$$\cos A = \frac{\text{adj.}}{\text{hyp.}} = \frac{23.1}{24.5} \approx 0.9429$$

$$\tan A = \frac{\text{opp.}}{\text{adj.}} = \frac{8.2}{23.1} \approx 0.3550$$

Reteaching with Practice

For use with pages 558–565

b. The length of the hypotenuse is 24.5. For $\angle B$, the length of the opposite side is 23.1 and the length of the adjacent side is 8.2.

$$\sin B = \frac{\text{opp.}}{\text{hyp.}} = \frac{23.1}{24.5} \approx 0.9429$$

$$\cos B = \frac{\text{adj.}}{\text{hyp.}} = \frac{8.2}{24.5} \approx 0.3347$$

$$\tan B = \frac{\text{opp.}}{\text{adj.}} = \frac{23.1}{8.2} \approx 2.8171$$

Exercises for Example 1

Find the sine, cosine, and tangent of $\angle A$.

1.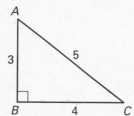

2.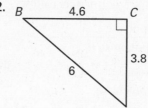

3.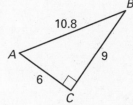

EXAMPLE 2 *Estimating a Distance*

It is known that a hill frequently used for sled riding has an angle of elevation of 30° at its bottom. If the length of a sledder's ride is 52.6 feet, estimate the height of the hill.

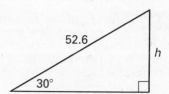

SOLUTION

Use the sine ratio for the 30° angle, because you have the value of the hypotenuse and you are looking for the value of the side opposite the 30° angle.

$$\sin 30° = \frac{h}{52.6}$$

$$h = (52.6) \cdot \sin 30° = (52.6) \cdot (0.5) = 26.3 \text{ feet}$$

Exercises for Example 2

4. In the sled-riding example, find the height of the hill if the angle of elevation of the hill is 42°.

5. If the angle of elevation from your position on the ground to the top of a building is 67° and you are standing 30 meters from the foot of the building, approximate the height of the building.

Quick Catch-Up for Absent Students

For use with pages 558–566

The items checked below were covered in class on (date missed) _____

Lesson 9.5: Trigonometric Ratios

_____ **Goal 1:** Find the sine, cosine, and the tangent of an acute angle. (pp. 558–560)

Material Covered:

_____ Example 1: Finding Trigonometric Ratios

_____ Example 2: Finding Trigonometric Ratios

_____ Student Help: Study Tip

_____ Example 3: Trigonometric Ratios for 45°

_____ Example 4: Trigonometric Ratios for 30°

_____ Student Help: Trigonometric Table

_____ Example 5: Using a Calculator

Vocabulary:

trigonometric ratio, p. 558   sine, p. 558

cosine, p. 558   tangent, p. 558

_____ **Goal 2:** Use trigonometric ratios to solve real-life problems. (p. 561)

Material Covered:

_____ Example 6: Indirect Measurement

_____ Example 7: Estimating a Distance

Vocabulary:

angle of elevation, p. 561

_____ Other (specify) _____

_____

Homework and Additional Learning Support

_____ Textbook (specify) _pp. 562–566_ _____

_____

_____ Internet: Extra Examples at www.mcdougallittell.com

_____ *Reteaching with Practice* worksheet (specify exercises)_____

_____ *Personal Student Tutor* for Lesson 9.5

NAME _____ DATE _____

Cooperative Learning Activity

For use with pages 558–566

GOAL **To find the height of a difficult to measure object using trigonometric ratios**

Materials: protractor, measuring tape or meter stick, calculator or trigonometric tables, drinking straw, plumb line, tape

Exploring Trigonometric Ratios

Trigonometric ratios can be used to determine the height of objects. Foresters use an instrument called a clinometer to measure the angle of elevation from the ground to the top of a tree. With this information, along with the distance to the tree, it is possible to use trigonometric ratios to solve for the height of the tree. This same technique can be used to find the height of many objects.

Instructions

❶ Construct the clinometer. Tape the string end of the plumb line to the vertex point of the protractor. Attach the straw to the straight edge of the protractor using the tape. Hold the device in such a way that when the top of the object is sighted through the viewing tube (straw), the plumb line crosses the angle measurements on the protractor (point C in the diagram below) forming an acute angle ($\angle AOC$). This angle is the complement of the angle of elevation (see diagram below).

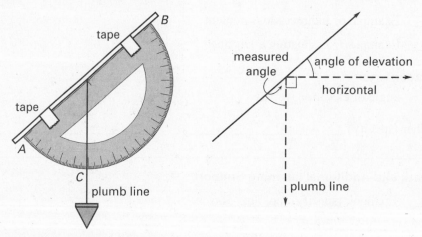

❷ Measure the eye height of the observer in the group.

❸ Locate the tall object that would be difficult to measure directly.

❹ Using the clinometer, measure the observer's viewing angle from the horizontal to the top of the object to be measured.

❺ Measure the distance to the base of the object.

Analyzing the Results

1. What is the height of the tall object?

2. Which trigonometric ratio was used to solve for the height?

3. Why is it impossible for the angle of elevation to be 90°?

NAME _____ DATE _____

Interdisciplinary Application

For use with pages 558–566

Leaning Tower of Pisa

HISTORY   The Leaning Tower of Pisa is located in Pisa, Italy and was built
over a period of about two centuries as a bell tower for a cathedral plaza. The
foundation work was started on August 9, 1173. Construction was stopped
many times, but the tower was finally completed around 1360. Historians
believe that the lean was evident, before construction stopped, at about the time
the third level was being added. Poor soil conditions and the weight of the
structure are the two main factors affecting the lean of the tower. In 1817,
measurements of the lean or inclination were taken using a plumb line. A plumb
line is a string with a weight (a plumb bob) attached. Suspended from the top
floor and stretching down to the ground, architects measured the distance from
the base of the foundation out to the plumb bob, and found that the tower was
leaning about 4.88° from the vertical. It was not until 1911 that precise
measurements of the tower's inclination began to be measured yearly using a
surveyor's tool called a theodolite. Currently, the tower measures 55.863 meters
high from the foundation and 55 meters from the ground, as shown in the
diagram below.

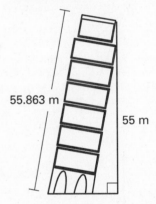

55.863 m

55 m

In Exercises 1–4, use the diagram above.

1. Find the distance from the base of the building out to the altitude.

2. Compute the angle that the tower makes with the ground.

3. How many degrees is the tower leaning from the vertical?

4. How many more degrees is the tower leaning than in 1811?

In Exercises 5 and 6, use the following information.

Upon completion of the tower in 1360, a measure of the lean was given as
1.63 meters. Assume that this is the distance from the base out to the altitude.

5. How many degrees was the tower leaning from the vertical in 1360?

6. What was the measurement from the top to the ground?

NAME _____ DATE _____

Challenge: Skills and Applications

For use with pages 558–566

1. Refer to the diagram. What is the length of the base $\overline{AB}$? Round to the nearest tenth.

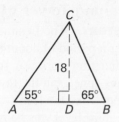

2. When the sun is shining at a 62° angle of elevation, a flagpole forms a shadow of length x feet. Later, the sun shines at an angle of 40°, and the shadow is 25 feet longer than before.

   a. Write two expressions for the height DE of the flagpole, in terms of x.

   b. How tall is the flagpole? Round to the nearest tenth of a foot.

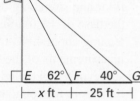

3. Refer to the diagram. Find JK and KL. Round decimals to the nearest tenth. (*Hint:* Draw an altitude.)

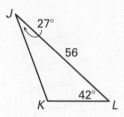

In Exercises 4–6, refer to the diagram.

4. Write an expression for $\tan x°$ and an expression for $\tan(90 - x)°$, in terms of a, b, and c. How is the tangent of an angle related to the tangent of the angle's complement?

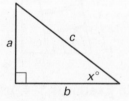

5. Write an expression for $(\sin x°)^2 + (\cos x°)^2$ in terms of a, b, and c. Then use the Pythagorean Theorem to simplify your expression.

6. If $\sin x° = 0.6$, what is the value of $\cos x°$?

7. Complete the following steps to evaluate $\sin 18°$.

   a. Show that $\triangle PRQ \sim \triangle QRT$.

   b. Use the similar triangles to find and solve a proportion involving x.

   c. Name an 18° angle in the diagram. What is the exact value of $\sin 18°$, in radical form?

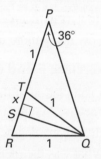

LESSON 9.5

NAME _____ DATE _____

Quiz 2

For use after Lessons 9.4 and 9.5

Sketch the figure that is described. Then find the requested information. Round decimals to the nearest tenth. *(Lesson 9.4)*

1. The side length of an equilateral triangle is 6 meters. Find the length of an altitude of the triangle.

2. The perimeter of a square is 32 inches. Find the length of the diagonal.

3. The side length of an equilateral triangle is 10 meters. Find the area of the triangle.

Find the value of each variable. Round decimals to the nearest tenth. *(Lesson 9.5)*

Answers

1. _____

2. _____

3. _____

4. _____

5. _____

6. _____

7. _____

4.
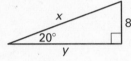

5.
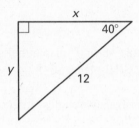

6.
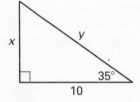

7. **Tower** You want to find the height of a tower used to transmit cellular phone calls. You stand 100 feet away from the tower and measure the angle of elevation to be 40°. How high is the tower? *(Lesson 9.5)*

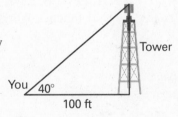

LESSON 9.6

Lesson Plan

1-day lesson (See *Pacing the Chapter,* TE pages 524C–524D)   **For use with pages 567–572**

GOALS   1. **Solve a right triangle.**
2. **Use right triangles to solve real-life problems.**

State/Local Objectives _____
_____

✓ **Check the items you wish to use for this lesson.**

STARTING OPTIONS

_____ Homework Check: TE page 562: Answer Transparencies
_____ Warm-Up or Daily Homework Quiz: TE pages 567 and 566, CRB page 84, or Transparencies

TEACHING OPTIONS

_____ Motivating the Lesson: TE page 568
_____ Lesson Opener (Application): CRB page 85 or Transparencies
_____ Technology Activity with Keystrokes: CRB pages 86–87
_____ Examples 1–3: SE pages 568–569
_____ Extra Examples: TE pages 568–569 or Transparencies; Internet
_____ Closure Question: TE page 569
_____ Guided Practice Exercises: SE page 570

APPLY/HOMEWORK

Homework Assignment

_____ Basic   12–32 even, 34–40, 42–45
_____ Average   12–32 even, 34–40, 42–45
_____ Advanced   12–32 even, 34–40, 42–46

Reteaching the Lesson

_____ Practice Masters: CRB pages 88–90 (Level A, Level B, Level C)
_____ Reteaching with Practice: CRB pages 91–92 or Practice Workbook with Examples
_____ Personal Student Tutor

Extending the Lesson

_____ Applications (Real-Life): CRB page 94
_____ Challenge: SE page 572; CRB page 95 or Internet

ASSESSMENT OPTIONS

_____ Checkpoint Exercises: TE pages 568–569 or Transparencies
_____ Daily Homework Quiz (9.6): TE page 572, CRB page 98, or Transparencies
_____ Standardized Test Practice: SE page 572; TE page 572; STP Workbook; Transparencies

Notes _____
_____
_____

TEACHER'S NAME _____ CLASS _____ ROOM _____ DATE _____

Lesson Plan for Block Scheduling

Half-day lesson (See *Pacing the Chapter*, TE pages 524C–524D)   **For use with pages 567–572**

 GOALS 1. **Solve a right triangle.**
2. **Use right triangles to solve real-life problems.**

State/Local Objectives _____

_____

_____

✓ **Check the items you wish to use for this lesson.**

STARTING OPTIONS

____ Homework Check: TE page 562: Answer Transparencies
____ Warm-Up or Daily Homework Quiz: TE pages 567 and
        566, CRB page 84, or Transparencies

CHAPTER PACING GUIDE	
Day	**Lesson**
1	Assess Ch. 8; 9.1 (all)
2	9.2 (all); 9.3 (begin)
3	9.3 (end); 9.4 (begin)
4	9.4 (end); 9.5 (begin)
5	9.5 (end); **9.6 (all)**
6	9.7 (all)
7	Review Ch. 9; Assess Ch. 9

TEACHING OPTIONS

____ Motivating the Lesson: TE page 568
____ Lesson Opener (Application): CRB page 85 or Transparencies
____ Technology Activity with Keystrokes: CRB pages 86–87
____ Examples 1–3: SE pages 568–569
____ Extra Examples: TE pages 568–569 or Transparencies; Internet
____ Closure Question: TE page 569
____ Guided Practice Exercises: SE page 570

APPLY/HOMEWORK

Homework Assignment **(See also the assignment for Lesson 9.5.)**
____ Block Schedule: 12–32 even, 34–40, 42–45

Reteaching the Lesson
____ Practice Masters: CRB pages 88–90 (Level A, Level B, Level C)
____ Reteaching with Practice: CRB pages 91–92 or Practice Workbook with Examples
____ Personal Student Tutor

Extending the Lesson
____ Applications (Real-Life): CRB page 94
____ Challenge: SE page 572; CRB page 95 or Internet

ASSESSMENT OPTIONS

____ Checkpoint Exercises: TE pages 568–569 or Transparencies
____ Daily Homework Quiz (9.6): TE page 572, CRB page 98, or Transparencies
____ Standardized Test Practice: SE page 572; TE page 572; STP Workbook; Transparencies

Notes _____

_____

_____

NAME _____ DATE _____

WARM-UP EXERCISES

For use before Lesson 9.6, pages 567–572

Find each indicated ratio. Express the answer to the nearest ten-thousandth.

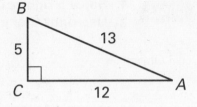

1. $\sin A$

2. $\cos A$

3. $\tan A$

4. $\sin B$

5. $\cos B$

6. $\tan B$

DAILY HOMEWORK QUIZ

For use after Lesson 9.5, pages 558–566

Find the sine, the cosine, and the tangent of the acute angles of the triangle. Express each value as a decimal rounded to four places.

1.

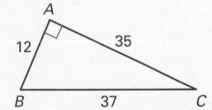

Use the diagram. Round decimals to the nearest tenth.

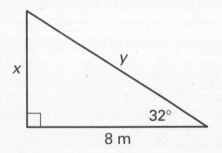

2. Find the value of each variable.

3. Find the area of the triangle.

NAME _____ DATE _____

Application Lesson Opener

For use with pages 567–572

The roof truss shown is
formed by two boards
joined at the top.

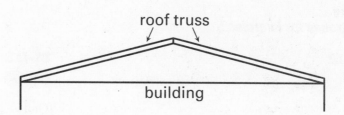

A carpenter uses
a framing square
to cut each board
for the truss.

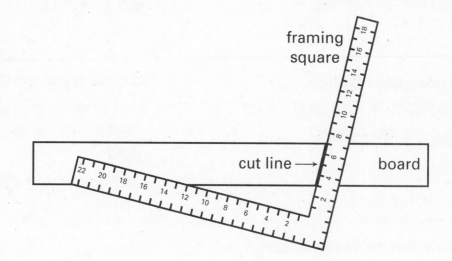

1. The carpenter places the framing square on a board at the
   inside measurements of 12 inches and 3 inches. Then the
   carpenter draws a line on the board along the inside of the
   framing square. This is the cut line. Sketch the right triangle
   formed by the framing square and the edge of the board. On
   the triangle you sketched, label the acute angle that is away
   from the cut line $\angle A$.

2. Angle A and the angle formed by the roof truss and the building have the same measure. You can use the measurements
   3 inches and 12 inches to find that angle measure. First write
   a trigonometric ratio using the triangle from Exercise 1.
   Then find $m\angle A$ using the Table of Trigonometric Ratios on
   page 845.

3. For another roof truss, the carpenter uses 5 inches and
   12 inches as inside measurements. Will this roof be *steeper* or
   flatter? Explain.

Technology Activity Keystrokes

For use with page 570

For use with page 570

Keystrokes for Exercise 5

TI-82

[MODE] [▼] [▼] [▶] [ENTER] [2nd]

[Quit]

[2nd] [TAN⁻¹] 5.4 [ENTER]

TI-83

[MODE] [▼] [▼] [▶] [ENTER] [2nd]

[Quit]

[2nd] [TAN⁻¹] [ )] 5.4 [ENTER]

SHARP EL-9600c

[2ndF] [SET UP] [B] 1 [2ndF] [QUIT]

[2ndF] [tan⁻¹] 5.4 [ENTER]

CASIO CFX-9850GA PLUS

From the main menu, choose RUN.

[SHIFT] [SET UP] [▼] [▼] [▼] [▼] [F1]

[EXIT]

[SHIFT] [tan⁻¹] 5.4 [EXE]

Geometry
Chapter 9 Resource Book

Lesson 9.6

NAME _____ DATE _____

Technology Activity Keystrokes

For use with page 570

Keystrokes for Exercise 17

TI-82

| MODE | ▼ | ▼ | ▶ | ENTER | 2nd |

[Quit]

| 2nd | [SIN⁻¹] 0.35 | ENTER |

TI-83

| MODE | ▼ | ▼ | ▶ | ENTER | 2nd |

[Quit]

| 2nd | [SIN⁻¹] |) | 0.35 | ENTER |

SHARP EL-9600c

| 2ndF | [SET UP] [B] 1 | 2ndF | [QUIT]
| 2ndF | [sin⁻¹] 0.35 | ENTER |

CASIO CFX-9850GA PLUS

From the main menu, choose RUN.

| SHIFT | [SET UP] | ▼ | ▼ | ▼ | ▼ | F1 |

| EXIT |

| SHIFT | [sin⁻¹] 0.35 | EXE |

Lesson 9.6

Geometry
Chapter 9 Resource Book
87

Practice A

For use with pages 567–572

Match the trigonometric expression with the correct ratio. Some ratios may be used more than once, and some may not be used at all.

1. $\sin A =$

2. $\cos A =$

3. $\tan A =$

4. $\sin B =$

5. $\cos B =$

6. $\tan B =$

A. $\dfrac{5}{13}$ **B.** $\dfrac{12}{13}$

C. $\dfrac{5}{12}$ **D.** $\dfrac{12}{5}$

E. $\dfrac{13}{12}$ **F.** $\dfrac{13}{5}$

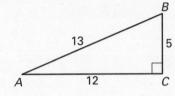

Use the diagram to find the indicated measurement. Round your answer to the nearest tenth.

7. CR

8. $m\angle T$

9. $m\angle C$

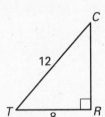

In Exercises 10–17, $\angle A$ is an acute angle. Use a calculator to approximate the measure of $\angle A$. Round to one decimal place.

10. $\sin A = 0.42$ **11.** $\tan A = 2.50$ **12.** $\cos A = 0.98$ **13.** $\sin A = 0.02$

14. $\cos A = 0.68$ **15.** $\tan A = 0.65$ **16.** $\sin A = 0.49$ **17.** $\tan A = 1.50$

Solve the right triangle. Round decimals to the nearest tenth.

18.

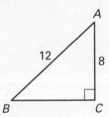

19.

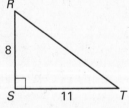

20.

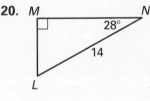

21.

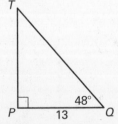

22.

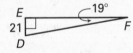

23.

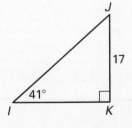

24. *Ladder*   You lean a 16 foot ladder against the wall. If the base is 4 feet from the wall, what angle does the ladder make with the ground?

NAME _____ DATE _____

Practice B

For use with pages 567–572

Use the diagram to find the indicated measurement. Round your answer to the nearest tenth.

1. MN

2. $m\angle M$

3. $m\angle N$

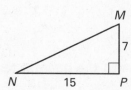

In Exercises 4–11, $\angle A$ is an acute angle. Use a calculator to approximate the measure of $\angle A$. Round to one decimal place.

4. $\sin A = 0.24$ 5. $\tan A = 1.73$ 6. $\cos A = 0.62$ 7. $\sin A = 0.08$

8. $\cos A = 0.94$ 9. $\tan A = 0.87$ 10. $\sin A = 0.38$ 11. $\tan A = 2.66$

Solve the right triangle. Round decimals to the nearest tenth.

12.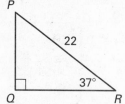

13.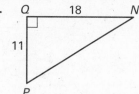

14.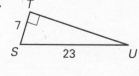

15.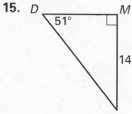

16.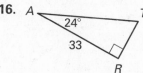

17.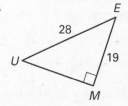

18. **Ramp**   A ramp was built by the loading dock. The height of the loading platform is 4 feet. Determine the length of the ramp if it makes a 32° angle with the ground.

19. **Office Buildings**   The angle of depression from the top of a 320 foot office building to the top of a 100 foot office building is 55°. How far apart are the two buildings?

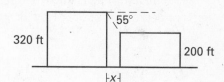

20. **Suspension Bridge**   Use the diagram to find the distance across the suspension bridge.

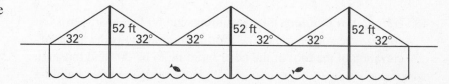

Practice C

For use with pages 567–572

In Exercises 1–8, ∠A is an acute angle. Use a calculator to approxi-mate the measure of ∠A. Round to one decimal place.

1. sin A = 0.85   **2.** tan A = 2.13   **3.** cos A = 0.87   **4.** sin A = 0.06

5. cos A = 0.15   **6.** tan A = 1.05   **7.** sin A = 0.42   **8.** tan A = 0.84

Solve the right triangle. Round decimals to the nearest tenth.

9.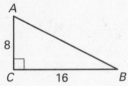

10.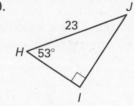

11.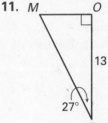

12.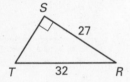

13.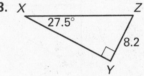

14.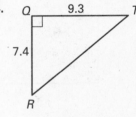

15.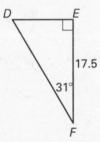

16.

17.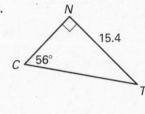

18. *Submarine*   A sonar operator on a ship detects a submarine at a distance of 400 meters and an angle of depression of 35°. How deep is the submarine?

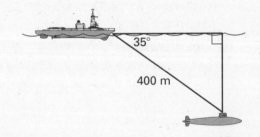

19. *Height of a Building*   Two buildings are 60 feet apart across a street. A person on top of the shorter building finds the angle of elevation of the roof of the taller building to be 20° and the angle of depression of its base to be 35°. How tall is the taller building to the nearest foot?

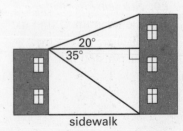

NAME _____ DATE _____

Reteaching with Practice

For use with pages 567–572

GOAL Solve a right triangle

VOCABULARY

To **solve a right triangle** means to determine the measures of all six parts.

EXAMPLE 1 *Solving a Right Triangle*

Solve the right triangle.

SOLUTION

Begin by using the Pythagorean Theorem to find the length of the missing side.

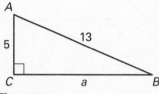

(hypotenuse)² = (leg)² + (leg)²	Pythagorean Theorem
$13^2 = a^2 + 5^2$	Substitute.
$169 = a^2 + 25$	Multiply.
$144 = a^2$	Subtract 25 from each side.
$12 = a$	Find the positive square root.

Then find the measure of $\angle B$.

$$\tan B = \frac{opp.}{adj.}$$

$$\tan B = \frac{5}{12}$$ Substitute.

$$m\angle B \approx 22.6° \quad \text{Use a calculator.}$$

Finally, because $\angle A$ and $\angle B$ are complements, you can write $m\angle A = 90° - m\angle B \approx 90° - 22.6° = 67.4°$.

The side lengths of $\triangle ABC$ are 5, 12, and 13. $\triangle ABC$ has one right angle and two acute angles whose measures are about 22.6° and 67.4°.

Exercises for Example 1

Solve the right triangle.

1.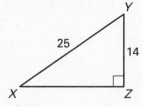

2.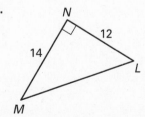

3.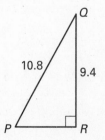

Reteaching with Practice

For use with pages 567–572

EXAMPLE 2 *Solving a Right Triangle*

Solve the right triangle.

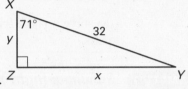

SOLUTION

Use trigonometric ratios to find the values of x and y.

$$\sin X = \frac{\text{opp.}}{\text{hyp.}}$$

$$\cos X = \frac{\text{adj.}}{\text{hyp.}}$$

$$\sin 71° = \frac{x}{32}$$

$$\cos 71° = \frac{y}{32}$$

$$32 \sin 71° = x$$

$$32 \cos 71° = y$$

$$32(0.9455) = x$$

$$32(0.3256) = y$$

$$30.3 \approx x$$

$$10.4 \approx y$$

Because $\angle X$ and $\angle Y$ are complements, you can write

$$m\angle Y = 90° - m\angle x = 90° - 71° = 19°.$$

The side lengths of the triangle are about 10.4, 30.3, and 32. The triangle has one right angle and two acute angles whose measures are 71° and 19°.

Exercises for Example 2

Solve the right triangle.

4.

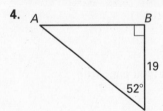

5.

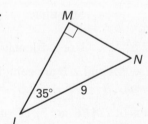

6.

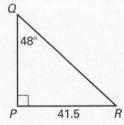

NAME _____ DATE _____

Quick Catch-Up for Absent Students

For use with pages 567–572

The items checked below were covered in class on (date missed) _____

Lesson 9.6: Solving Right Triangles

____ **Goal 1:** Solve a right triangle. (pp. 567–568)

Material Covered:

____ Activity: Finding Angles in Right Triangles

____ Example 1: Solving a Right Triangle

____ Student Help: Study Tip

____ Example 2: Solving a Right Triangle

Vocabulary:

solve a right triangle, p. 567

____ **Goal 2:** Use right triangles to solve real-life problems. (p. 569)

Material Covered:

____ Example 3: Solving a Right Triangle

____ Other (specify) _____

_____

Homework and Additional Learning Support

____ Textbook (specify) _pp. 570–572_____

_____

____ Internet: Extra Examples at www.mcdougallittel.com

____ *Reteaching with Practice* worksheet (specify exercises)_____

____ *Personal Student Tutor* for Lesson 9.6

NAME _____   DATE _____

Real-Life Application:
When Will I Ever Use This?

For use with pages 567–572

Hubble Space Telescope

The Hubble Space Telescope (HST) is a telescope in orbit around Earth. Being above the atmosphere allows it to receive images and signals from the stars and planets without distortion. The HST was set up on April 25, 1990 from the space shuttle Discovery. Since then it has captured stunning images of planets and stars. It has discovered galaxies as far away as 7 billion light years. A light year is the distance light travels in a year.

One of the uses of the HST is to help scientists with Astrometry, which is the science of measuring the distance of stars and planets from Earth and each other. The HST is a radio telescope that gives accurate measurements of star distances from Earth but trigonometry must be used to calculate the distance from star to star. The illustration below shows stars B and C and their relative position to Earth located at point A. The stars are 20 degrees apart in the sky and the HST has discovered that star B is 50 units from Earth and star C is 35 units from Earth. One unit is 10,000 light years.

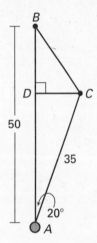

In Exercises 1–5, use the diagram above.

1. Use a trigonometric ratio to find *CD*. Round to the nearest tenth.

2. Use a trigonometric ratio to find *AD*. Round to the nearest tenth.

3. Find *BD* to the nearest tenth.

4. What method can you use to find *BC* knowing *CD* and *BD*?

5. Find *BC* in light years.

NAME _____ DATE _____

Challenge: Skills and Applications

For use with pages 567–572

1. Refer to the diagram.

   a. Write an expression for h using $\angle A$.

   b. Write an expression for h using $\angle B$.

   c. Show that $\dfrac{\sin A}{a} = \dfrac{\sin B}{b}$.

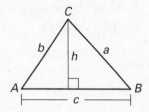

In Exercise 1, you proved part of the *Law of Sines*. Given $\triangle ABC$ with side lengths a, b, and c, the Law of Sines states that $\dfrac{\sin A}{a} = \dfrac{\sin B}{b} = \dfrac{\sin C}{c}$.

In Exercises 2–4, use the Law of Sines to solve the triangle. That is, find the unknown measures of sides or angles. Round lengths of sides to the nearest tenth.

2.

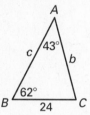

3.

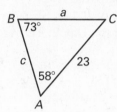

4.

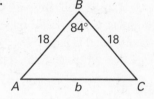

5. Refer to the diagram.

   a. Use $\triangle ACD$ to write expressions for $p^2 + q^2$ and for p in terms of b and $\angle A$.

   b. Find a^2 in terms of c, p, and q. Give your answer in expanded (not factored) form.

   c. Use your results from parts (a) and (b) to find an expression for a^2 in terms of b, c, and $\angle A$.

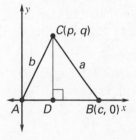

In Exercise 5, you proved part of the *Law of Cosines*. Given $\triangle ABC$ with side lengths a, b, and c, the Law of Cosines states that $a^2 = b^2 + c^2 - 2bc \cos A$, $b^2 = a^2 + c^2 - 2ac \cos B$, and $c^2 = a^2 + b^2 - 2ab \cos C$.

In Exercises 6–8, use the Law of Cosines, the Law of Sines, or both to solve the triangle.

6.

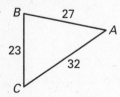

7.

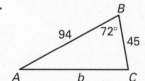

8.

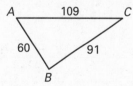

Lesson Plan

2-day lesson (See *Pacing the Chapter,* TE pages 524C–524D)   For use with pages 573–580

 GOALS   1. **Find the magnitude and the direction of a vector.**
2. **Add vectors.**

State/Local Objectives _____

_____

✓ **Check the items you wish to use for this lesson.**

STARTING OPTIONS
____ Homework Check: TE page 570: Answer Transparencies
____ Warm-Up or Daily Homework Quiz: TE pages 573 and 572, CRB page 98, or Transparencies

TEACHING OPTIONS
____ Lesson Opener (Application): CRB page 99 or Transparencies
____ Technology Activity with Keystrokes: CRB pages 100–102
____ Examples: Day 1: 1–3, SE pages 573–574; Day 2: 4–5, SE page 575
____ Extra Examples: Day 1: TE page 574 or Transp.; Day 2: TE page 575 or Transp.
____ Closure Question: TE page 575
____ Guided Practice: SE page 576   Day 1: Exs. 1–8; Day 2: Ex. 9

APPLY/HOMEWORK
Homework Assignment
____ Basic   Day 1: 10–30; Day 2: 31–45, 48, 53–60; Quiz 3: 1–7
____ Average   Day 1: 10–30; Day 2: 31–48, 53–60; Quiz 3: 1–7
____ Advanced   Day 1: 10–30; Day 2: 31–60; Quiz 3: 1–7

Reteaching the Lesson
____ Practice Masters: CRB pages 103–105 (Level A, Level B, Level C)
____ Reteaching with Practice: CRB pages 106–107 or Practice Workbook with Examples
____ Personal Student Tutor

Extending the Lesson
____ Applications (Interdisciplinary): CRB page 109
____ Challenge: SE page 579; CRB page 110 or Internet

ASSESSMENT OPTIONS
____ Checkpoint Exercises: Day 1: TE page 574 or Transp.; Day 2: TE page 575 or Transp.
____ Daily Homework Quiz (9.7): TE page 580, or Transparencies
____ Standardized Test Practice: SE page 579; TE page 580; STP Workbook; Transparencies
____ Quiz (9.6–9.7): SE page 580

Notes _____

_____

_____

TEACHER'S NAME _____ CLASS _____ ROOM _____ DATE _____

Lesson Plan for Block Scheduling

1-day lesson (See *Pacing the Chapter*, TE pages 524C–524D)   For use with pages 573–580

 GOALS
1. **Find the magnitude and the direction of a vector.**
2. **Add vectors.**

State/Local Objectives _____

_____

_____

✓ **Check the items you wish to use for this lesson.**

STARTING OPTIONS

____ Homework Check: TE page 570: Answer Transparencies

____ Warm-Up or Daily Homework Quiz: TE pages 573 and
        572, CRB page 98, or Transparencies

TEACHING OPTIONS

____ Lesson Opener (Application): CRB page 99 or Transparencies

____ Technology Activity with Keystrokes: CRB pages 100–102

____ Examples 1–5: SE pages 573–575

____ Extra Examples: TE pages 574–575 or Transparencies

____ Closure Question: TE page 575

____ Guided Practice Exercises: SE page 576

APPLY/HOMEWORK
Homework Assignment

____ Block Schedule: 10–48, 53–60; Quiz 3: 1–7

Reteaching the Lesson

____ Practice Masters: CRB pages 103–105 (Level A, Level B, Level C)

____ Reteaching with Practice: CRB pages 106–107 or Practice Workbook with Examples

____ Personal Student Tutor

Extending the Lesson

____ Applications (Interdisciplinary): CRB page 109

____ Challenge: SE page 579; CRB page 110 or Internet

ASSESSMENT OPTIONS

____ Checkpoint Exercises: TE pages 574–575 or Transparencies

____ Daily Homework Quiz (9.7): TE page 580, or Transparencies

____ Standardized Test Practice: SE page 579; TE page 580; STP Workbook; Transparencies

____ Quiz (9.6–9.7): SE page 580

Notes _____

_____

_____

CHAPTER PACING GUIDE	
Day	**Lesson**
1	Assess Ch. 8; 9.1 (all)
2	9.2 (all); 9.3 (begin)
3	9.3 (end); 9.4 (begin)
4	9.4 (end); 9.5 (begin)
5	9.5 (end); 9.6 (all)
6	**9.7 (all)**
7	Review Ch. 9; Assess Ch. 9

Lesson 9.7

WARM-UP EXERCISES

For use before Lesson 9.7, pages 573–580

**Find the distance between each pair of points.
Leave answers in simplest radical form.**

1. $(0, 0), (3, 4)$ **2.** $(1, -4), (5, -1)$

3. $(-1, 3), (4, 2)$ **4.** $(0, 0), (4, 4)$

5. $(7, 2), (2, 0)$

DAILY HOMEWORK QUIZ

For use after Lesson 9.6, pages 567–572

**Use a calculator to approximate the measure of acute
∠A to the nearest tenth of a degree.**

1. $\sin A = 0.25$

2. $\cos A = 0.38$

**Solve the right triangle. Round decimals to the nearest
tenth.**

3.

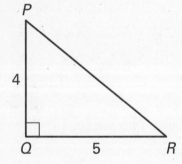

4.

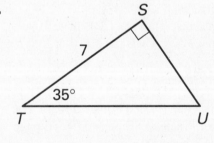

NAME _____ DATE _____

Application Lesson Opener

For use with pages 573–580

Set up: You will need: • ruler

Car A approaches an intersection at a speed of 15 miles per hour, crosses the intersection, and then, after the intersection, accelerates to a speed of 30 miles per hour. Car B approaches a different intersection at a speed of 5 miles per hour, makes a turn, and then accelerates to a speed of 25 miles per hour. In the diagram below, each car is shown before and after the intersection.

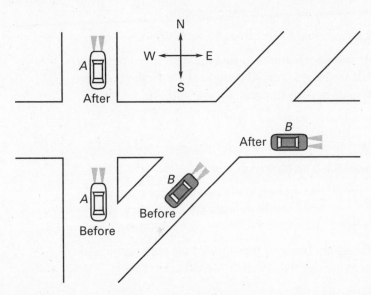

1. The velocity of a moving car can be represented by a vector, which describes both the speed and the direction of the moving car. Give the speed and direction of Car A *before* it reaches the intersection and *after* it has crossed the intersection and accelerated. Do the same for Car B. (Use north, south, east, and west for directions.)

2. The vectors at the right represent the velocity of each car *before* reaching the intersection. How do the speeds of Car A and Car B compare during that part of their trips? How is that shown by the two vectors?

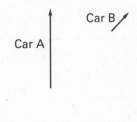

3. Use a ruler to draw vectors that represent the velocity of each car *after* it has crossed the intersection and accelerated.

Technology Activity

For use with pages 573–580

GOAL **To use geometry software to solve a real-life problem**

A boat leaves a dock represented on the coordinate axes as (2, 2) and is headed east at 10 knots. A wind is blowing from the south at 4 knots. Find the boat's new speed and the angle at which the boat has been blown off course.

Activity

❶ Turn on the coordinate axes and the grid. Adjust the grid scale to accommodate the problem situation.

❷ Construct the boat's vector and the wind's vector.

❸ Use the parallelogram rule to construct the sum vector.

❹ Use appropriate software features to find the new speed (take into account the scale of the grid) and the angle at which the boat has been blown off course. Also find the slope of the sum vector.

Exercises

1. What is the boat's new speed?

2. At what angle was the boat blown off course?

3. If the wind's speed increased to 6 knots and the boat's speed remained at 10 knots, what would the new speed and new direction of the boat be?

4. What is the relationship between the slope of the sum vector and the angle at which the boat has been blown off course?

Technology Activity Keystrokes

For use with pages 573–580

TI-92

1. Turn on the coordinate axes and the grid. Adjust the grid scale to accommodate the problem situation.

   [F8] 9 (Set Coordinate Axes to RECTANGULAR and Grid to ON.)

   [ENTER]

   [F1] 1 (Place cursor on a point on x-axis; drag the point until the scale changes from 0.5 to 2.)

2. Construct the boat's vector and the wind's vector.

   [F2] 7 (Place cursor on (2, 2).) [ENTER] (Move cursor to (12, 2).) [ENTER]

   [F2] 7 (Place cursor on (2, 2).) [ENTER] (Move cursor to (2, 6).) [ENTER]

3. Use the parallelogram rule to construct the sum vector.

   [F2] 7 (Place cursor on (2, 6).) [ENTER] (Move cursor to (12, 6).) [ENTER]

   [F2] 7 (Place cursor on (12, 2).) [ENTER] (Move cursor to (12, 6).)

   [ENTER]

   [F2] 7 (Place cursor on (2, 2).) [ENTER] (Move cursor to (12, 6).) [ENTER]

4. Find the new speed (take into account the scale of the grid).

   [F6] 1 (Place cursor on a horizontal vector.) [ENTER] (Move cursor to sum vector.) [ENTER]

   [F6] 6 [( ] [10] [÷] (Use cursor to highlight the value of the horizontal vector.) [ENTER] [ )] (Highlight the value of the sum vector.)

   [ENTER] [ENTER] (The result will appear on the screen.)

   Find the angle at which the boat has been blown off course ([F6] 3).

   Find the slope of the sum vector.

   [F6] 4 (Place cursor on sum vector.) [ENTER]

NAME _____ DATE _____

Technology Activity Keystrokes

For use with pages 573–580

SKETCHPAD

1. Turn on the coordinate axes and the grid. Choose **Snap to Grid** from the **Graph** menu. To change grid size, use the translate selection arrow tool to drag point *B*.

2. Construct the boat's vector and the wind's vector. Choose segment from the straightedge tools. Construct a segment from (2, 2) to (12, 2). Construct a segment from (2, 2) to (12, 6).

3. Use the parallelogram rule to construct the sum vector. Construct a segment from (2, 6) to (12, 6). Construct a segment from (12, 2) to (12, 6). Construct a segment from (2, 2) to (12, 6).

4. Find the new speed (take into account the scale of the grid). Choose **Calculate** from the **Measure** menu. Click [( ] , enter 10, click [÷] , click the horizontal vector, click [ )] click [*] , click the value of the sum vector, and click OK.

   Find the angle at which the boat has been blown off course. Use the selection arrow tool, to select the points that make up the angle.

   Choose **Angle** from the **Measure** menu.

   Find the slope of the sum vector. Use the selection arrow tool to highlight the sum vector. Choose **Slope** from the **Measure** menu.

NAME _____ DATE _____

Practice A

For use with pages 573–580

Match the vector with the correct component form of the vector.

1.

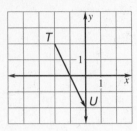

2.

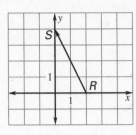

3.

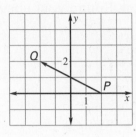

A. $\langle -2, 4 \rangle$

B. $\langle -4, 2 \rangle$

C. $\langle 2, -4 \rangle$

Write the vector in component form. Find the magnitude of the vector. Round your answer to the nearest tenth.

4.

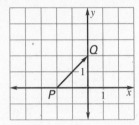

5.

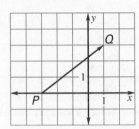

6.

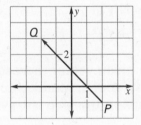

Draw vector $\overrightarrow{PQ}$ in a coordinate plane. Write the component form of the vector and find its magnitude. Round your answer to the nearest tenth.

7. $P(0, 0), Q(3, 4)$

8. $P(0, 0), Q(5, 8)$

9. $P(2, 1), Q(7, 8)$

10. $P(-2, 4), Q(1, -5)$

The given vector represents the velocity of a ship at sea. Find the ship's speed, rounded to the nearest mile per hour. Then find the direction the ship is traveling relative to the given direction.

11. Find direction relative to east.

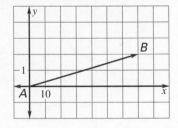

12. Find direction relative to east.

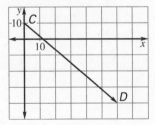

Geometry
Chapter 9 Resource Book

103

Write the vector in component form. Find the magnitude of the vector. Round your answer to the nearest tenth.

1.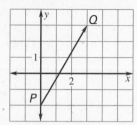

2.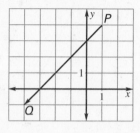

3.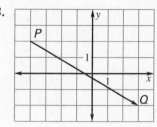

Draw vector $\overrightarrow{PQ}$ in a coordinate plane. Write the component form of the vector and find its magnitude. Round your answer to the nearest tenth.

4. $P(0, 0), Q(5, 2)$

5. $P(2, 5), Q(6, 1)$

6. $P(-4, 2), Q(2, 0)$

7. $P(2, -3), Q(-1, 3)$

The given vector represents the velocity of a ship at sea. Find the ship's speed, rounded to the nearest mile per hour. Then find the direction the ship is traveling relative to the given direction.

8. Find direction relative to east.

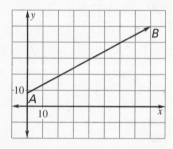

9. Find direction relative to west.

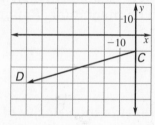

In Exercises 10–13, use the diagram shown at the right.

10. Which vectors are parallel?

11. Which vectors have the same direction?

12. Which vectors are equal?

13. Name two vectors that have the same magnitude but different directions.

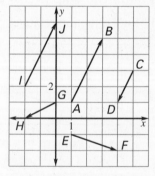

Practice C

For use with pages 573–580

Write the vector in component form. Find the magnitude of the vector. Round your answer to the nearest tenth.

1.
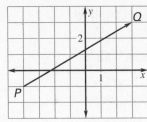

2.
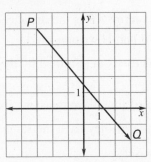

3.
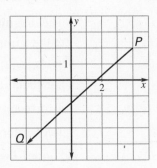

The given vector represents the velocity of a ship at sea. Find the ship's speed, rounded to the nearest mile per hour. Then find the direction the ship is traveling relative to the given direction.

4. Find direction relative to west.

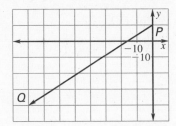

5. Find direction relative to east.

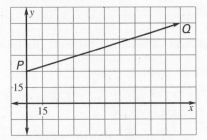

Copy the vectors $\vec{u}$ and $\vec{v}$. Write the component form of each vector. Then find the sum $\vec{u} + \vec{v}$ and draw the vector $\vec{u} + \vec{v}$.

6.
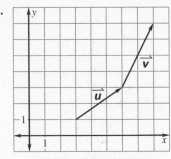

7.
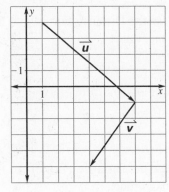

Let $\vec{u} = \langle 2, -3 \rangle$, $\vec{v} = \langle 3, -2 \rangle$, and $\vec{z} = \langle 4, -1 \rangle$. Find the given sum.

8. $\vec{u} + \vec{v}$

9. $\vec{v} + \vec{z}$

10. $\vec{z} + \vec{u}$

Reteaching with Practice

For use with pages 573–579

GOAL **Find the magnitude and the direction of a vector and add vectors**

VOCABULARY

The **magnitude of a vector** $\overrightarrow{AB}$ is the distance from the initial point A to the terminal point B and is written $|\overrightarrow{AB}|$.

The **direction of a vector** is determined by the angle it makes with a horizontal line.

Two vectors are **equal** if they have the same magnitude and direction.

Two vectors are **parallel** if they have the same or opposite directions.

Sum of Two Vectors

The sum of $\vec{u} = \langle a_1, b_1 \rangle$ and $\vec{v} = \langle a_2, b_2 \rangle$ is

$\vec{u} + \vec{v} = \langle a_1 + a_2, b_1 + b_2 \rangle$.

EXAMPLE 1 *Finding the Magnitude of a Vector*

Points P and Q are the initial and terminal points of the vector $\overrightarrow{PQ}$.

Draw $\overrightarrow{PQ}$ in a coordinate plane. Write the component form of the vector and find its magnitude.

a. $P(1, 2)$, $Q(5, 5)$ **b.** $P(0, 4)$, $Q(-2, -4)$

SOLUTION

a. Component form $= \langle x_2 - x_1, y_2 - y_1 \rangle$

$$\overrightarrow{PQ} = \langle 5 - 1, 5 - 2 \rangle$$
$$= \langle 4, 3 \rangle$$

Use the Distance Formula to find the magnitude.

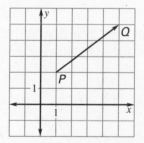

$$|\overrightarrow{PQ}| = \sqrt{(5 - 1)^2 + (5 - 2)^2} = \sqrt{25} = 5$$

b. Component form $= \langle x_2 - x_1, y_2 - y_1 \rangle$

$$\overrightarrow{PQ} = \langle -2 - 0, -4 - 4 \rangle$$
$$= \langle -2, -8 \rangle$$

Use the Distance Formula to find the magnitude.

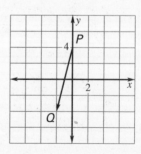

$$|\overrightarrow{PQ}| = \sqrt{(-2 - 0)^2 + (-4 - 4)^2} = \sqrt{68} \approx 8.2$$

Exercises for Example 1

Draw $\overrightarrow{PQ}$ in a coordinate plane. Write the component form of the vector and find its magnitude.

1. $P(3, 2)$, $Q(1, 9)$ **2.** $P(-2, 1)$, $Q(0, -5)$

3. $P(3, 8)$, $Q(-1, 10)$ **4.** $P(-4, -11)$, $Q(0, 2)$

NAME _____ DATE _____

Reteaching with Practice

For use with pages 573–579

EXAMPLE 2 *Describing the Direction of a Vector*

The vector $\overrightarrow{AB}$ depicts the velocity of a moving vehicle. The scale on each axis is in kilometers per hour. Find the (a) speed of the vehicle and (b) direction it is traveling relative to east.

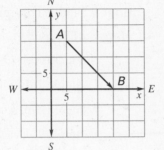

SOLUTION

a. The magnitude of the vector $\overrightarrow{AB}$ represents the vehicle's speed. Use the Distance Formula.

$$|\overrightarrow{AB}| = \sqrt{(20 - 5)^2 + (0 - 15)^2}$$
$$= \sqrt{15^2 + 15^2}$$
$$\approx 21.2$$

The speed of the vehicle is about 21.2 kilometers per hour.

b. The tangent of the angle formed by the vector and a line drawn parallel to the *x*-axis at point *A* is $-\dfrac{15}{15} = -1$. Use a calculator to find the angle measure.

[2nd] [TAN] $-1 = -45°$

The vehicle is traveling in a direction 45° south of east.

Exercises for Example 2

In Exercises 5–7, find the vehicle's magnitude and direction if points *A* and *B* are as given.

5. $A(0, 0), B(6, 7)$ **6.** $A(-2, 4), B(3, -1)$ **7.** $A(2, 4), B(-3, -1)$

EXAMPLE 3 *Finding the Sum of Two Vectors*

Let $\vec{u} = \langle -4, 2 \rangle$ and $\vec{v} = \langle 3, 1 \rangle$ Write the component form of the sum $\vec{u} + \vec{v}$.

SOLUTION

To find the sum vector $\vec{u} + \vec{v}$, add the horizontal components and add the vertical components of $\vec{u}$ and $\vec{v}$.

$$\vec{u} + \vec{v} = \langle -4 + 3, 2 + 1 \rangle = \langle -1, 3 \rangle$$

Exercises for Example 3

For the given vectors $\vec{u}$ and $\vec{v}$, find the component form of the sum $\vec{u} + \vec{v}$.

8. $\vec{u} = \langle 0, 8 \rangle$ and $\vec{v} = \langle -3, 5 \rangle$ **9.** $\vec{u} = \langle -2, -7 \rangle$ and $\vec{v} = \langle 2, 10 \rangle$

10. $\vec{u} = \langle 3, 12 \rangle$ and $\vec{v} = \langle -3, -12 \rangle$

Lesson 9.7

NAME _____ DATE _____

Quick Catch-Up for Absent Students

For use with pages 573–580

The items checked below were covered in class on (date missed) _____

Lesson 9.7: Vectors

_____ **Goal 1:** Find the magnitude and the direction of a vector. (pp. 573–574)

Material Covered:

_____ Student Help: Look Back

_____ Example 1: Finding the Magnitude of a Vector

_____ Student Help: Look Back

_____ Example 2: Describing the Direction of a Vector

_____ Example 3: Identifying Equal and Parallel Vectors

Vocabulary:

magnitude of a vector, p. 573   direction of a vector, p. 574

equal, p. 574   parallel, p. 574

_____ **Goal 2:** Add vectors. (p. 575)

Material Covered:

_____ Student Help: Study Tip

_____ Example 4: Finding the Sum of Two Vectors

_____ Example 5: Velocity of a Jet

Vocabulary:

sum, p. 575

_____ Other (specify) _____

_____

Homework and Additional Learning Support

_____ Textbook (specify) _pp. 576–580_____

_____

_____ *Reteaching with Practice* worksheet (specify exercises)_____

_____ *Personal Student Tutor* for Lesson 9.7

Geometry
Chapter 9 Resource Book

NAME _____ DATE _____

Interdisciplinary Application

For use with pages 573–580

Projectile Motion

PHYSICS   Vectors can be used to model projectile motion. Projectile motion occurs when an object is projected into the air. When an object with mass is flying through the air, its motion has both a vertical and horizontal component. The horizontal velocity of the object remains constant while the vertical velocity accelerates or decelerates due to gravity. You can also use vectors to find how high and how far an object is projected (horizontal and vertical displacement).

If an object were projected with an initial velocity of 88 feet per second, you would use the following equations to find the vertical and horizontal displacement of the object. These two formulas work together to simulate the path of a projectile as it is launched and as it is pulled back down to Earth by the force of gravity.

Horizontal displacement: $x = 88t\cos\theta$

Vertical displacement: $y = 88t\sin\theta - 16t^2$

In Exercises 1–5, use the equations above and the following information.

You are building a rocket for a physics class competition. You are competing to see which team's rocket will go the farthest down the football field once it is launched in the air. The restrictions limit the initial velocity of each launch to 88 feet per second (60 miles per hour). Team 1 launches their rocket at an angle of 45° with the ground, and Team 2 launches their rocket at an angle of 20° with the ground.

1. What are the simplified vertical and horizontal displacement equations for Team 1?

2. What are the simplified vertical and horizontal displacement equations for Team 2?

3. Which team's rocket is furthest down the field after 0.5 seconds? After 1 second?

4. When the rockets hit the ground, the distance from the launch to where they land is measured and a winner is determined. Each rocket goes up in the air vertically, but gravity pulls each back to Earth. Find how long each rocket is in the air by setting y equal to 0 and solving for t.

5. Which team won the competition and by how much?

Challenge: Skills and Applications

For use with pages 573–580

Vectors can be subtracted and multiplied, as well as added.

Let $\vec{u} = \langle a_1, b_1 \rangle$ and $\vec{v} = \langle a_2, b_2 \rangle$. The following definitions apply.

Difference of two vectors: $\quad\quad\quad \vec{u} - \vec{v} = \langle a_1 - a_2, b_1 - b_2 \rangle$

Product of a number and a vector: $\quad k\vec{u} = \langle ka_1, kb_1 \rangle$

Dot product of two vectors: $\quad\quad\quad \vec{u} \cdot \vec{v} = a_1 a_2 + b_1 b_2$

In particular, the dot product of two vectors may not be what you would expect. Notice that it is a *number*, not a vector.

In Exercises 1–5, show algebraically that the property applies to vectors.

Example: Commutative property of addition:

$$\vec{u} + \vec{v} = \langle a_1 + a_2, b_1 + b_2 \rangle = \langle a_2 + a_1, b_2 + b_1 \rangle = \vec{v} + \vec{u}$$

So, $\vec{u} + \vec{v} = \vec{v} + \vec{u}$.

1. Associative property of addition: $(\vec{u} + \vec{v}) + \vec{w} = \vec{u} + (\vec{v} + \vec{w})$

2. Commutative property for the dot product: $\vec{u} \cdot \vec{v} = \vec{v} \cdot \vec{u}$

3. Distributive property for the product of a number and a vector:
   $k(\vec{u} + \vec{v}) = k\vec{u} + k\vec{v}$

4. Distributive property for the dot product: $\vec{u} \cdot (\vec{v} + \vec{w}) = \vec{u} \cdot \vec{v} + \vec{u} \cdot \vec{w}$

5. An associative property of multiplication: $(k\vec{u}) \cdot \vec{v} = k(\vec{u} \cdot \vec{v})$

In Exercises 6–13, evaluate each expression, where $\vec{u} = \langle 5, 3 \rangle$, $\vec{v} = \langle 3, -2 \rangle$, $\vec{w} = \langle 1, 0 \rangle$, and $\vec{z} = \langle -4, 6 \rangle$.

6. $\vec{u} - \vec{v}$ $\quad\quad\quad\quad\quad\quad\quad\quad$ 7. $7\vec{u}$

8. $\vec{v} \cdot \vec{z}$ $\quad\quad\quad\quad\quad\quad\quad\quad$ 9. $2\vec{v} + 5\vec{w}$

10. $5\vec{z} - \vec{v}$ $\quad\quad\quad\quad\quad\quad\quad$ 11. $(3\vec{v}) \cdot \vec{w}$

12. $\vec{v} \cdot (2\vec{z} - \vec{u})$ $\quad\quad\quad\quad\quad$ 13. $(\vec{u} + \vec{w}) \cdot (\vec{v} - \vec{z})$

NAME _____ DATE _____

Chapter Review Games and Activities

For use after Chapter 9

The answer to each question below is a positive integer. Put the answers to the first six questions in the circles on the first triangle in such a way that the sum of the numbers along any side is 14. Do the same for the last six questions and the second triangle.

Triangle 1 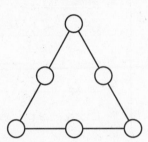   Triangle 2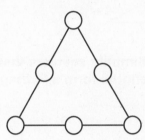

1. The length of the shorter leg of a 30°-60°-90° triangle is 3. What is the length of the hypotenuse?

2. The length of one leg of a right triangle is 1 and the length of the hypotenuse is $\sqrt{50}$. What is the length of the other leg?

3. The length of one leg of 45°-45°-90° triangle is $\sqrt{2}$. What is the length of the hypotenuse?

4. The length of one leg of a right triangle is $\sqrt{17}$ and the length of the other leg is $\sqrt{8}$. What is the length of the hypotenuse?

5. The length of the longer leg of a 30°-60°-90° triangle is $\sqrt{108}$. What is the length of the shorter leg?

6. The length of the hypotenuse of a right triangle is 18 and the sine of $\angle A$ is $\frac{1}{6}$. What is the length of the side opposite $\angle A$?

7. The length of one leg of a right triangle is equal to the length of the other leg. What is the tangent of the angle between the hypotenuse and one of the other sides?

8. A vector has an initial point of $(2, 4)$ and a terminal point of $(5, 8)$. What is the magnitude of this vector?

9. What is the largest number the sine of an angle could be?

10. What is the first component of the sum of $\langle -1, -2 \rangle$ and $\langle 5, -3 \rangle$?

11. A vector has an initial point of $(1, -5)$ and a terminal point of $(2, 3)$. What is the second component of this vector?

12. In a right triangle, the altitude from the right angle to the hypotenuse divides the hypotenuse into two segments, one of length 10 and the other of length $\frac{5}{2}$. What is the length of the altitude?

Review and Assess

NAME _____ DATE _____

Chapter Test A

For use after Chapter 9

Complete and solve the proportion.

1. $\dfrac{x}{8} = \dfrac{?}{16}$

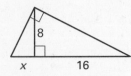

2. $\dfrac{4}{x} = \dfrac{x}{?}$

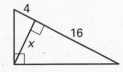

3. $\dfrac{x}{3} = \dfrac{?}{x}$

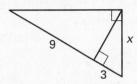

Find the unknown side length. Simplify answers that are radicals. Tell whether the side lengths form a Pythagorean triple.

4.

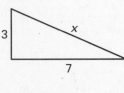

5.

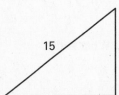

6.

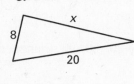

Decide whether the numbers can represent the side lengths of a triangle. If they can, classify the triangle as *right*, *acute*, or *obtuse*.

7. 6, 8, 10

8. 3, 4, 6

9. 6, 2, 5

10. 5.4, 3.8, 10

11. 1, 2, 3

12. 1.6, 3.0, 3.4

Find the value of each variable. Write answers in simplest radical form.

13.

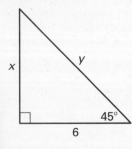

14.

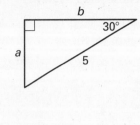

15.

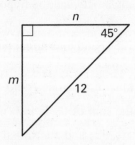

Answers

1. a. _____
   b. _____
2. a. _____
   b. _____
3. a. _____
   b. _____
4. _____ ; _____
5. _____ ; _____
6. _____ ; _____
7. _____
8. _____
9. _____
10. _____
11. _____
12. _____
13. _____
14. _____
15. _____

NAME _____ DATE _____

Chapter Test A

For use after Chapter 9

Find the sine, the cosine, and the tangent of the acute angle A of the triangle. Express each value as a decimal rounded to four places.

16.

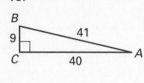

17.

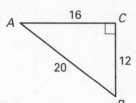

18.

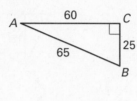

Use trigonometric ratios to find the value of each variable. Round decimals to the nearest tenth.

19.

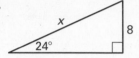

20.

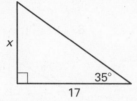

Draw vector $\overrightarrow{PQ}$ in a coordinate plane. Write the component form of the vector and find its magnitude. Round decimals to the nearest tenth.

21. $P(0, 0), Q(2, 4)$

22. $P(1, 0), Q(5, 2)$

23. $P(-1, -2), Q(-3, -5)$

24. $P(1, -1), Q(-4, 3)$

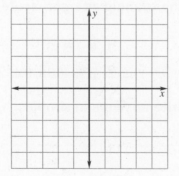

Let $\vec{u} = \langle 2, 5 \rangle, \ \vec{v} = \langle 3, 7 \rangle, \vec{w} = \langle -3, -7 \rangle,$ and $\vec{z} = \langle 5, 2 \rangle$. Find the given sum.

25. $\vec{v} + \vec{w}$

26. $\vec{u} + \vec{z}$

27. $\vec{v} + \vec{z}$

28. $\vec{w} + \vec{z}$

29. $\vec{u} + \vec{v}$

30. $\vec{u} + \vec{w}$

16. a. _____
   b. _____
   c. _____

17. a. _____
   b. _____
   c. _____

18. a. _____
   b. _____
   c. _____

19. _____

20. _____

21. _____

22. _____

23. _____

24. _____

25. _____

26. _____

27. _____

28. _____

29. _____

30. _____

Review and Assess

NAME _____ DATE _____

Chapter Test B

For use after Chapter 9

Complete and solve the proportion.

1. $\dfrac{x}{12} = \dfrac{?}{8}$

2. $\dfrac{?}{x-4} = \dfrac{4}{6}$

3. $\dfrac{x}{16} = \dfrac{16}{?}$

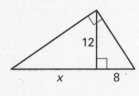

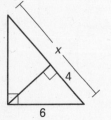

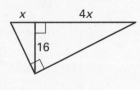

Find the unknown side length. Simplify answers that are radicals. Tell whether the side lengths form a Pythagorean triple.

4.

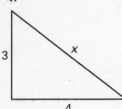

5.

6.

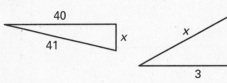

Decide whether the numbers can represent the side lengths of a triangle. If they can, classify the triangle as *right*, *acute*, or *obtuse*.

7. 7, 7, 4

8. 8, 9, 15

9. 9, 12, 13

10. 6, 8, 10

11. 1, 2, 8

12. 0.27, 0.36, 0.45

Find the value of each variable. Write answers in simplest radical form.

13.

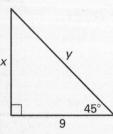

14.

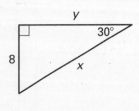

15.

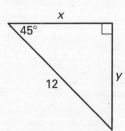

Answers

1. a. _____
   b. _____
2. a. _____
   b. _____
3. _____
   _____
4. _____ ; _____
5. _____ ; _____
6. _____ ; _____
7. _____
8. _____
9. _____
10. _____
11. _____
12. _____
13. _____
14. _____
15. _____

Review and Assess

NAME _____ DATE _____

Chapter Test B

For use after Chapter 9

Find the sine, the cosine, and the tangent of the acute angle A of the triangle. Express each value as a decimal rounded to four places.

16.

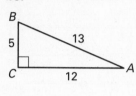

17.

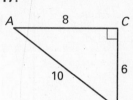

18.

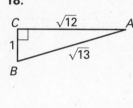

16. a. _____
   b. _____
   c. _____

17. a. _____
   b. _____
   c. _____

18. a. _____
   b. _____
   c. _____

Use trigonometric ratios to find the value of each variable. Round decimals to the nearest tenth.

19.

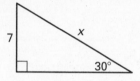

20.

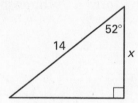

19. _____

20. _____

21. _____

_____

22. _____

Draw vector $\overrightarrow{PQ}$ in a coordinate plane. Write the component form of the vector and find its magnitude. Round decimals to the nearest tenth.

21. $P(0, 0)$, $Q(5, 3)$

22. $P(-1, 0)$, $Q(4, -5)$

23. $P(2, -5)$, $Q(-2, -5)$

24. $P(-3, -3)$, $Q(5, 0)$

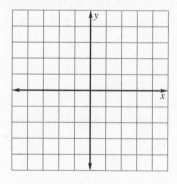

23. _____

_____

24. _____

_____

25. _____

26. _____

27. _____

28. _____

Let $\vec{u} = \langle 1, 6 \rangle$, $\vec{v} = \langle -1, 6 \rangle$, $\vec{w} = \langle 3, 5 \rangle$, and $\vec{z} = \langle -5, -7 \rangle$. Find the given sum.

29. _____

30. _____

25. $\vec{v} + \vec{w}$ **26.** $\vec{u} + \vec{z}$

27. $\vec{v} + \vec{z}$ **28.** $\vec{w} + \vec{z}$

29. $\vec{u} + \vec{v}$ **30.** $\vec{u} + \vec{w}$

Review and Assess

NAME _____   DATE _____

Chapter Test C

For use after Chapter 9

Complete and solve the proportion.

1. $\dfrac{x}{7} = \dfrac{?}{3}$

2. $\dfrac{?}{x} = \dfrac{3}{8}$

3. $\dfrac{x}{10} = \dfrac{10}{?}$

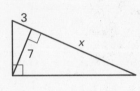

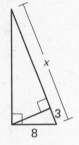

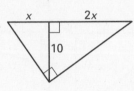

Find the unknown side length. Simplify answers that are radicals. Tell whether the side lengths form a Pythagorean triple.

4.

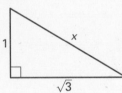

5.

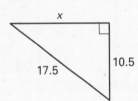

6.

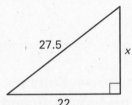

Decide whether the numbers can represent the side lengths of a triangle. If they can, classify the triangle as *right*, *acute*, or *obtuse*.

7. 6, 8, 10

8. 2, 6, 5

9. 1, 2, 8

10. 0.27, 0.36, 0.45

11. 2.5, 1.5, 4

12. 12, 16, 20

Find the value of each variable. Write answers in simplest radical form.

13.

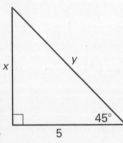

14.

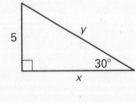

15.

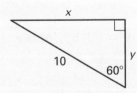

1.	a. _____
	b. _____
2.	a. _____
	b. _____
3.	_____
	_____
4.	_____ ; _____
5.	_____ ; _____
6.	_____ ; _____
7.	_____
8.	_____
9.	_____
10.	_____
11.	_____
12.	_____
13.	_____
14.	_____
15.	_____

Find the sine, the cosine, and the tangent of the acute angle _A_ of the triangle. Express each value as a decimal rounded to four places.

16.

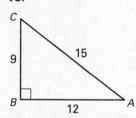

17.

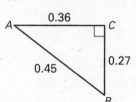

18.

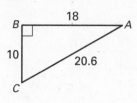

16. a. _____
b. _____
c. _____
17. a. _____
b. _____
c. _____
18. a. _____
b. _____
c. _____

Use trigonometric ratios to find the value of each variable. Round decimals to the nearest tenth.

19.

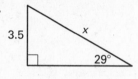

20.

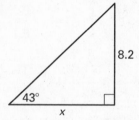

19. _____
20. _____
21. _____
_____
_____
22. _____
_____

Draw vector $\overrightarrow{PQ}$ in a coordinate plane. Write the component form of the vector and find its magnitude. Round decimals to the nearest tenth.

21. $P(0, 0)$, $Q(-3, -8)$

22. $P(-3, 5)$, $Q(-5, 3)$

23. $P(-2, -9)$, $Q(3, 5)$

24. $P(2, 6)$, $Q(-1, 5)$

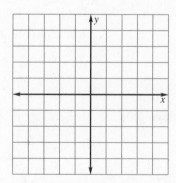

23. _____
_____
_____
24. _____
_____
25. _____
26. _____
27. _____
28. _____
29. _____
30. _____

Let $\vec{u} = \langle 0.5, 2.75 \rangle$, $\vec{v} = \langle -3, -7 \rangle$, $\vec{w} = \langle 2.25, -5 \rangle$, and

$\vec{z} = \langle -4, 3 \rangle$. Find the given sum.

25. $\vec{v} + \vec{w}$

26. $\vec{u} + \vec{z}$

27. $\vec{v} + \vec{z}$

28. $\vec{w} + \vec{z}$

29. $\vec{u} + \vec{v}$

30. $\vec{u} + \vec{w}$

Review and Assess

NAME _____ DATE _____

SAT/ACT Chapter Test

For use after Chapter 9

1. Use the diagram to find the values of x and y.

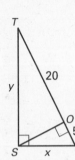

- **A** $x = 5\sqrt{5}, y = 10\sqrt{5}$
- **B** $x = 5\sqrt{5}, y = 20\sqrt{5}$
- **C** $x = 5\sqrt{25}, y = 10\sqrt{5}$
- **D** $x = 5\sqrt{25}, y = 20\sqrt{5}$
- **E** $x = 5\sqrt{15}, y = 20\sqrt{5}$

2. In the diagram below, what is the measure of $\angle A$ to the nearest tenth of a degree?

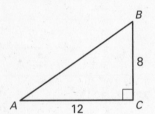

- **A** $41.8°$
- **B** $48.2°$
- **C** $33.7°$
- **D** $1°$
- **E** $42°$

3. Which set of numbers can represent the side lengths of an obtuse triangle?

- **A** 12, 16, 20
- **B** 8, 14, 17
- **C** 1, 2, 1
- **D** 3, 4, 5
- **E** 3.5, 3.5, 3.5

4. Points $A(5, 2)$ and $B(8, 7)$ are the initial and the terminal points of $\overrightarrow{AB}$. Find the magnitude of $\overrightarrow{AB}$.

- **A** $\langle 3, 5 \rangle$
- **B** $\langle 3, 1 \rangle$
- **C** $2\sqrt{6}$
- **D** $\langle 5, 3 \rangle$
- **E** $\sqrt{34}$

5. Let $\vec{v} = \langle -3, y \rangle$ and $\vec{w} = \langle x, 8 \rangle$. If $\vec{v} + \vec{w} = \langle 1, 3 \rangle$, what are the values of x and y?

- **A** $x = -4, y = 11$
- **B** $x = 4, y = 11$
- **C** $x = -4, y = -5$
- **D** $x = 4, y = -5$
- **E** $x = -4, y = 5$

6. Let the numbers represent the lengths of the sides of a triangle. Which of the triangles are right triangles?

- **A** 5, 8, 13
- **B** 27, 36, 45
- **C** 1, 2, 3
- **D** 7.5, 8.5, 10.5
- **E** 18, 24, 31

7. The length of a diagonal of a square is 20 inches. What is its perimeter?

- **A** $40\sqrt{2}$ in.
- **B** $20\sqrt{2}$ in.
- **C** $30\sqrt{2}$ in.
- **D** 20 in.
- **E** $10\sqrt{2}$ in.

8. The base of an isosceles triangle is 21 centimeters long. The altitude to the base is 9 centimeters long. What is the approximate measure of a base angle of the triangle?

- **A** $60°$
- **B** $49.4°$
- **C** $40.6°$
- **D** $31°$
- **E** $42°$

9. Find the area of $\square RSTU$.

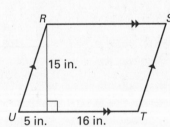

- **A** 253 in.^{2}
- **B** 332 in.^{2}
- **C** 277.5 in.^{2}
- **D** 240 in.^{2}
- **E** 315 in.^{2}

10. Using the figure in Exercise 9, find the perimeter of $\square RSTU$.

- **A** 72.5 in.
- **B** 73.6 in.
- **C** 72 in.
- **D** 73 in.
- **E** 71.8 in.

NAME _____   DATE _____

Alternative Assessment and Math Journal

For use after Chapter 9

JOURNAL   **1.** Use the diagram below. List six pairs of similar triangles.
If $NK = 22$ and $KL = 11$, then find JK.

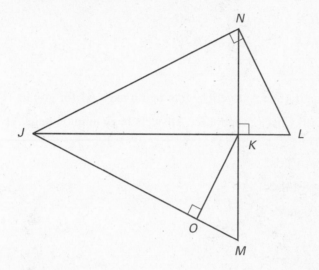

MULTI-STEP   **2.** While golfing you drive a ball into the woods. The ball ricochets
PROBLEM   off a tree at a 90° angle and lands in the fairway. Your shot can be
modeled by $\vec{v} = \langle 12, 14 \rangle$. The ricochet off the tree can be modeled
by $\vec{u} = \langle -7, 6 \rangle$.

   a. Draw the vectors in a coordinate plane.

   b. Find the sum vector of $\vec{v}$ and $\vec{u}$. Label the vector $\vec{s}$. Draw $\vec{s}$ in the
   same coordinate plane.

   c. Find the distance in yards the ball traveled before it hit the tree.

   d. Find the distance in yards the ball traveled after it hit the tree.

   e. How far in yards are you going to walk to get to your ball?

   f. Find the measure of the angle formed by the path of your shot
   before it hit the tree and the path you walked to get to your ball.

3. *Critical Thinking*   Your partner does the same thing. Your partner's
shot ricochets off a tree at a 90° angle and lands in the fairway. The
measure of the angle formed by the path of your partner's shot
before it hit the tree and the path your partner walks to get to the ball
is 30°. Your partner walked 50 yards to reach the ball.

   a. How far did the ball travel before it hit the tree?

   b. How far did the ball travel after it hit the tree?

4. *Writing*   Write a paragraph proof. A diagram will help.

   Given: In $\triangle EFG$, $\angle FGE$ is a right angle.

   Prove: $a^2 + b^2 = c^2$

Review and Assess

Alternative Assessment Rubric

For use after Chapter 9

JOURNAL SOLUTION

1. Complete answers should include:

$\triangle NLK \sim \triangle JLN$ $\triangle KMO \sim \triangle JMK$

$\triangle JNK \sim \triangle JLN$ $\triangle JKO \sim \triangle JMK$

$\triangle NLK \sim \triangle JNK$ $\triangle KMO \sim \triangle JKO$

$JK = 44$

MULTI-STEP PROBLEM SOLUTION

2. **a.** Check graph. (See Answers beginning on page A1 for graphs.)

    b. $\vec{s} = \langle 5, 20 \rangle$. Check graph. (See Answers beginning on page A1 for graphs.)

    c. About 18.4 yards

    d. About 9.4 yards

    e. About 20.6 yards

    f. 27.1°

3. **a.** About 43.3 yards

    b. 25 yards

4. Answers may vary; the final statement should prove $a^2 + b^2 = c^2$.

MULTI-STEP PROBLEM RUBRIC

4   Students answer all parts of the problem correctly, showing work in a step-by-step manner. Students' graphs are correct and clearly labeled. Students' proof is correct and follows a logical order.

3   Students complete all parts of the problem. Students' graphs are complete, but may have a labeling error. Students' proof is complete, but may contain an incorrect reason.

2   Students complete the problem. Students' graphs are complete, but may contain an incorrect vector and incorrect labels. Students' proof is complete, but may contain incorrect statements and reasons.

1   Students' answers are incomplete. Students' graphs are incomplete. Students' proof is incomplete and the final statement is not proven.

Review and Assess

Project: Proving a Conjecture

For use with Chapter 9

OBJECTIVE **Find and compare the costs of two pieces of property.**

MATERIALS graph paper, ruler, and calculator

INVESTIGATION Your parents want your help in analyzing two plots of land they are considering purchasing. Your parents have measured the size of each property by walking in a pattern, making right and left turns of 90°. The boundaries of the property are the lines connecting the stakes. You now have the following descriptions:

Lot 1: Begin at stake 1, go right 180 ft to stake 2, turn right, go 120 ft, turn right and go 60 ft to stake 3, go straight 60 ft, turn left, go 60 ft, turn left and go 60 ft to stake 4, turn right, go 60 ft, turn right and go 60 ft to stake 5, go straight 120 ft, turn right and go 90 ft to stake 6, turn left, go 30 ft, turn right and go 90 ft to stake 7, turn right, go 90 ft, turn left and go 60 ft, returning to stake 1.

Lot 2: Begin at stake 1, go right 100 ft to stake 2, turn right, go 50 ft, turn right and go 25 ft to stake 3, turn left, go 25 ft, turn left and go 75 ft to stake 4, turn right, go 100 ft, turn right and go 100 ft to stake 5, go straight 75 ft to stake 6, turn right, go 75 ft, turn left and go 50 ft to stake 7, turn right go 100 ft, turn right and go 75 ft, returning to stake 1.

1. Draw a diagram for each piece of property showing your parents' path. Identify the scale you are using and mark the location of each stake.
2. Draw another diagram showing the actual outline of each piece of property. Use the same scale and mark the location of each stake.
3. Find the perimeter and area of each piece of property. Explain your methods and show your calculations.
4. Determine the cost of fencing the properties. Fencing costs $6.75 per foot.
5. Determine the cost of the taxes for the first year. Taxes for Lot 1 are $683.54 per acre. Lot 2 is in another town where taxes are $725.38 per acre. (*Hint:* First determine how many square feet are in an acre.)
6. Based on the cost of the taxes and fencing, describe any advantages of purchasing each piece of property.
7. Which piece of property would you recommend that your parents purchase if Lot 1 is selling for $12,500 and Lot 2 is selling for $11,800? Explain.

PRESENT YOUR RESULTS Your report should contain all of the diagrams. You should explain your methods and show your calculations for determining the area, perimeter, fencing costs, and taxes. Include an examination of any right triangles you formed in calculating area or perimeter. Are any of them similar? Are there any special right triangles? If so, explain how you can use those relationships in finding or checking measurements. Discuss the advantages and disadvantages of each property. Make and support a recommendation to your parents about which piece of property to purchase.

Project: Teacher's Notes

For use with Chapter 9

GOALS
- Follow a directional pattern.
- Create and use a scale drawing.
- Use the Pythagorean Theorem and special right triangle's.
- Determine the total area of irregularly shaped property.

MANAGING THE PROJECT

Remind students that their drawings must end up at Stake 1. When diagrams are complete, ask students how to find lengths between stakes. In addition to using the Pythagorean Theorem, students may find some lengths by recognizing a special right triangle or a triangle similar to a 3-4-5 right triangle.

Ask students how they can find the area of an irregular shape. Expect a variety of approaches. You might want to review area formulas, in particular the formula for the area of a trapezoid. For accurate calculation of tax charges, acreage should be determined to the nearest hundredth of an acre. Students can research measurements for an acre or you may tell them ($43{,}560$ ft^2 = 1 acre).

Concluding the Project   Have a discussion with students asking how many chose each lot. You could even have students who made different choices debate the rationale for their choice.

RUBRIC   **The following rubric can be used to assess student work.**

4   All diagrams and calculations for area, perimeter, fencing costs, and taxes are accurate and are neatly presented in an organized manner. The explanation of calculation methods is thorough and shows an excellent grasp of how to apply the right triangle techniques of the chapter and how to break a polygon into smaller parts for which formulas are known. The student thoughtfully compares the properties and makes a well-supported purchasing suggestion.

3   The student's diagrams and calculations for area, perimeter, fencing costs, and taxes are complete and mostly accurate, but are not neatly organized or may contain minor errors. The explanation of calculation methods includes some appropriate use of right triangle techniques and shows a basic grasp of breaking a polygon into smaller parts for which formulas are known, but more explanation would be helpful. The student describes some advantages or disadvantages and makes a suggestion for purchasing property which is somewhat supported.

2   The student completes the diagrams and calculations for area, perimeter, fencing costs, and taxes, but work contains errors. An explanation of calculation methods is included, but may be unclear or show some misunderstandings. The student makes a suggestion for purchasing property, but not enough support is provided.

1   Diagrams and calculations for area, perimeter, fencing costs, and taxes are not complete or contain many errors. The explanation of calculation methods is missing or shows a lack of understanding of basic ideas. The recommendation for purchasing property is missing or lacks explanation.

NAME _____ DATE _____

Cumulative Review

For use after Chapters 1–9

Find the length of *AB*. Write your answer to the nearest tenth. (1.3)

1. $A(5, 2)$, $B(-7, 4)$

2. $A(-3, 1)$, $B(5, 9)$

Write the (a) inverse and (b) converse of the statement. (2.1)

3. If $m\angle 1 = 110°$, then $\angle 1$ is an obtuse angle.

4. If the sun is shining, then it is not raining.

State the postulate or theorem you would use to prove $l \parallel m$. (3.3)

5.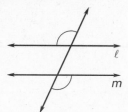

6.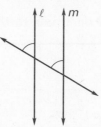

Prove using a two-column format. (4.3, 4.4, 4.6)

7. **Given:** $\overline{MN} \cong \overline{PN}$

   O is a midpoint of $\overline{MP}$.

   Prove: $\triangle MON \cong \triangle PON$

8. **Given:** $\overline{GH} \parallel \overline{KJ}$

   $\overline{LG} \cong \overline{LK}$

   Prove: $\angle H \cong \angle K$

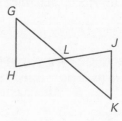

Complete the following, using the given information. (5.4)

$XZ = 4$, $BC = 10$, perimeter of $\triangle XYZ = 16$

9. $AB =$ ____?____

10. Perimeter of $\triangle ABC =$ ____?____ .

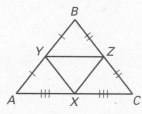

Give the most descriptive name of the figure. (6.6)

11.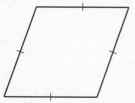

12.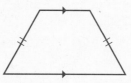

Give the coordinates of the image of $\overrightarrow{AB}$ using the translation vector described. (7.4)

13. $\langle -2, 3 \rangle$

14. $\langle 3, 1 \rangle$

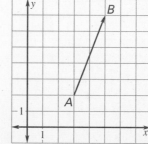

Geometry
Chapter 9 Resource Book

123

The two polygons are similar. Find *x* and *y*. (8.3)

15.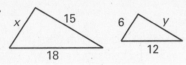

16.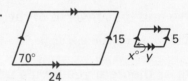

Find the given length. Round decimals to the nearest tenth. (9.1)

17. Find *ML*.

18. Find *KL*.

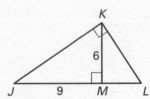

The numbers below represent the sides of a triangle. Classify the triangle as *right*, *acute* or *obtuse*. (9.3)

19. 5, 8, 10

20. 5, 12, 13

Find the value of the given variables. Write answers in simplest radical form if possible. (9.4, 9.5)

21.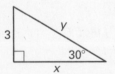

22.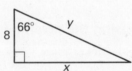

Solve the right triangle. Round your answers to the nearest tenth. (9.6)

23.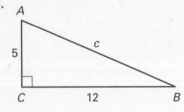

24.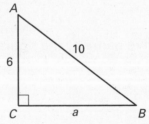

Write the component form of the vector. Find its magnitude, round your answer to the nearest tenth. (9.7)

25. $\overrightarrow{AB}$

26. $\overrightarrow{PQ}$

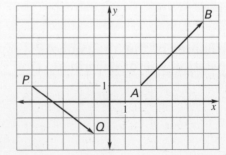

Review and Assess

ANSWERS

Chapter Support

Parent Guide

9.1: about 44 ft **9.2:** about 33 cm² **9.3:** 6, 8, and 10 spaces **9.4:** $15\sqrt{3}$ or almost 26 ft
9.5: 0.0600; 0.0599; 0.9982
9.6: $RS \approx 24.4$ in., $QS \approx 25.4$ in., $m\angle Q = 74°$
9.7: $\langle -5, 3 \rangle$, about 5.8

Prerequisite Skills Review

1. $6\sqrt{2}$ **2.** $4\sqrt{5}$ **3.** $3\sqrt{5}$ **4.** $4\sqrt{2}$
5. $3\sqrt{10}$ **6.** $2\sqrt{13}$ **7.** 10 **8.** -6 **9.** 4
10. 9 **11.** 5 **12.** 1 **13.** $\langle 3, 6 \rangle$ **14.** $\langle -7, 5 \rangle$
15. $\langle 2, 2 \rangle$ **16.** $\langle -5, -1 \rangle$ **17.** $\langle -10, -3 \rangle$
18. $\langle 3, 4 \rangle$

Strategies for Reading Mathematics

1. 1.414 **2.** 0.0349 **3.** 0.0524
4. Look for 0.0349 in the Tangent column of the table. Then move across to the left-hand column to find the angle measure; 2°.

Lesson 9.1

Warm-Up Exercises

1. 3 **2.** 6; -6 **3.** 3 **4.** $4\sqrt{3}$; $-4\sqrt{3}$
5. 20

Daily Homework Quiz

1. The dilation has center C and scale factor $\frac{3}{5}$; $x = 10.8$, $y = 9$
2. $X'(15, 5)$, $Y'(30, 20)$, $Z'(40, -5)$
3. $P'(-4, 2)$, $Q'(2, 2)$, $R'(2, -4)$, $S'(-4, -4)$

Lesson Opener

Allow 15 minutes.
1. *Sample answer:*

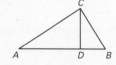

---

2. *Sample answer:*
$\triangle CDB \sim \triangle ADC \sim \triangle ACB$

3. *Sample answer:* Yes

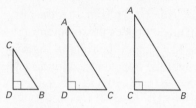

Practice A

1. $\overline{AB}, \overline{BC}$ **2.** $\overline{AC}$ **3.** 45°, 45° **4.** $\overline{DE}, \overline{EF}$
5. $\overline{DF}$ **6.** No; $\triangle DEF$ has a right angle; therefore, the remaining angles must be complementary so that the angles sum to 180°. **7.** $4\sqrt{5}$
8. 15 **9.** $7\sqrt{6}$
10.

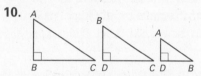

11. $\triangle ABC \sim \triangle BDC \sim \triangle ADB$ **12.** 4; 2
13. 12; $6\sqrt{3}$ **14.** 5; $2\sqrt{10}$
15. $\triangle ACB \sim \triangle ADC \sim \triangle CDB$; CD
16. $\triangle GFE \sim \triangle FHE \sim \triangle GHF$; GH
17. $\triangle IKJ \sim \triangle ILK \sim \triangle KLJ$; IK

Practice B

1. 12; 18 **2.** 5; $5\sqrt{3}$ **3.** 20; $6\sqrt{5}$
4. $\triangle PQR \sim \triangle PSQ \sim \triangle QSR$; $4\sqrt{5}$
5. $\triangle TUV \sim \triangle TWU \sim \triangle UWV$; $2\sqrt{65}$
6. $\triangle XYZ \sim \triangle XOY \sim \triangle YOZ$; $\frac{81}{5}$ **7.** 3 **8.** $4\sqrt{3}$
9. $\frac{25}{6}$ **10.** 1. Given 2. If an altitude is drawn to the hypotenuse, then the two triangles formed are $\sim$ to the original and to each other. 3. Given
4. Corresponding Angles Postulate 5. Reflexive Property of $\cong$ 6. AA Similarity Postulate
7. Transitive Property

Practice C

1. 9 **2.** 25 **3.** 6 **4.** 9 **5.** 4 **6.** 2
7. ≈ 46.48 in. **8.** ≈ 37.95 in. **9.** ≈ 29.39 in.

10. $\angle ACB \cong \angle ADC \cong \angle CDB$,
$\angle ABC \cong \angle CBD \cong \angle ACD$,
$\angle BAC \cong \angle BCD \cong \angle CAD$,
$\angle ADF \cong \angle FDC \cong \angle CDE \cong \angle EDB$

11. $\triangle ABC \sim \triangle CBD \sim \triangle ACD$ **12.** true

13. no   **14.** false   **15.** Using Thm. 9.3, the altitude from the right angle divides the hypotenuse into two segments. Each leg of the right triangle is the geometric mean of the hypotenuse and the segment of the hypotenuse that is adjacent to the leg.

$$\frac{DC}{BC} = \frac{BC}{AC} \Longrightarrow \frac{4}{BC} = \frac{BC}{6} \Longrightarrow BC^2 = 24 \Longrightarrow$$

$BC = 2\sqrt{6}$ **16.** First find JM by the Pythagorean Thm. $JM^2 + MK^2 = KJ^2 \Longrightarrow$
$JM^2 + 9 = 25 \Longrightarrow JM = 4$. Next, using Thm. 9.3, the altitude from the right angle divides the hypotenuse into two segments. Each leg of the right triangle is the geometric mean of the hypotenuse and the segment of the hypotenuse that is adjacent to the leg. Also, let $ML = x$, so $JL = x + 4$, then

$$\frac{MJ}{KJ} = \frac{KJ}{JL} \Longrightarrow \frac{4}{5} = \frac{5}{x + 4} \Longrightarrow 4x + 16 = 25 \Longrightarrow$$

$x = \dfrac{9}{4}$ so $JL = 4 + \dfrac{9}{4} = \dfrac{25}{4}$.

Reteaching with Additional Practice

1. 2.4 units   **2.** 1 unit   **3.** $\approx$ 1.5 units
4. 6.5 units   **5.** 11.5 units   **6.** 11.4 units

Interdisciplinary Application

1. a. $\dfrac{OU}{OV} = \dfrac{OV}{OW}$ **b.** $\dfrac{OV}{OW} = \dfrac{OW}{OX}$

c. $\dfrac{OW}{OX} = \dfrac{OX}{OY}$ **d.** $\dfrac{OX}{OY} = \dfrac{OY}{OZ}$

2. $\overline{OV}$ **3.** The length of each segment is the geometric mean of the lengths of the segments before and after it.

4. $\dfrac{ZX}{XY} = \dfrac{XY}{OX}; \dfrac{ZX}{YZ} = \dfrac{YZ}{OZ}$

Challenge: Skills and Applications

1. *Sample answer:* Construct the perpendicular bisector of $\overline{AB}$ to find the midpoint E of $\overline{AB}$. Construct a circle centered at E containing B. Then $\overline{AB}$ is a diameter of the circle. Construct a perpendicular to $\overline{AB}$ that contains D. Let C be either of the two points where the perpendicular bisector intersects the circle. Construct $\overline{CD}$; the length of $\overline{CD}$ is the geometric mean of AD and BD since $\overline{CD}$ is the altitude to the hypotenuse of right $\triangle ABC$.   **2. a.** *Sample answer:* $\triangle ABC$ can be inscribed in a circle; since $\triangle ABC$ is a right triangle, the hypotenuse $\overline{AB}$ is a diameter of the circle. Since $\overline{CE}$ is a median, E is the midpoint of $\overline{AB}$, so E is the center of the circle. Therefore, $\overline{AE}, \overline{BE}$, and $\overline{CE}$ are all radii of the same circle, and so they are congruent. Now, the arithmetic mean of AD and BD is $\dfrac{AD + BD}{2} = \dfrac{AB}{2} = AE = CE$.

b. *Sample answer:* Since $\overline{CE}$ is the hypotenuse of right triangle $\triangle CDE$, the arithmetic mean CE is greater than the geometric mean CD.

c. *Sample answer:* Given distinct positive numbers x and y, choose points A, B, and D such that D is between A and B, AD is equal to the larger of x or y, and BD is equal to the smaller of x or y. By following the procedure in the answer to Exercise 1, a figure like the one shown in Exercise 2 can be constructed. The argument in the answer to part (b) shows that the arithmetic mean of x and y is greater than the geometric mean of x and y.

3. *Sample answer:* Since $\dfrac{AB}{CB} = \dfrac{CB}{DB}$, $(BC)^2 = (CB)^2 = AB \cdot BD$.

Since $\dfrac{AB}{AC} = \dfrac{AC}{AD}$, $(AC)^2 = AB \cdot AD$.

Therefore, $\dfrac{(AC)^2}{(BC)^2} = \dfrac{AB \cdot AD}{AB \cdot BD} = \dfrac{AD}{BD}$.

4. $AC = x\sqrt{x^2 + y^2}$, $BC = y\sqrt{x^2 + y^2}$, $CD = xy$

5. 6   **6.** 12   **7.** $\frac{5}{3}$ **8.** 3   **9.** 7   **10.** 14

Lesson 9.2

Warm-Up Exercises

1. 8.5; -8.5 **2.** 6; -6 **3.** 9.2; -9.2
4. 8.4; -8.4 **5.** 15; -15

Daily Homework Quiz

1. $\triangle ADC \sim \triangle ABD \sim \triangle DBC$ **2.** BD
3. $\frac{64}{15} \approx 4.3$

Lesson 9.2 *continued*

Lesson Opener

Allow 15 minutes.

To complete the proof, students should write the following Steps 6–10.

6. $\dfrac{\dfrac{a^2}{b}}{DB} = \dfrac{a}{c}$ (Substitution) **7.** $DB = \dfrac{ac}{b}$ (Cross product prop.; multiplication prop. of equality)

8. Area of $\triangle BCD$ + Area of $\triangle BCA$ = Area of $\triangle DBA$ (Sum of areas)

9. $\dfrac{1}{2} \cdot a \cdot \dfrac{a^2}{b} + \dfrac{1}{2} \cdot a \cdot b = \dfrac{1}{2} \cdot \dfrac{ac}{b} \cdot c$

(Area of $a\triangle = \dfrac{1}{2} \cdot$ base $\cdot$ height)

10. $a^2 + b^2 = c^2$

$\left(\text{Multiplication prop. of equality; multiply by } \dfrac{2b}{a}.\right)$

Technology Activity

1. a. 24.5 **b.** 144.5 **c.** 84.5

2. $\dfrac{1}{2}(a + b)(a + b) = \dfrac{1}{2}ab + \dfrac{1}{2}cc + \dfrac{1}{2}ab$

$(a + b)(a + b) = ab + cc + ab$

$a^2 + 2ab + b^2 = c^2 + 2ab$

$a^2 + b^2 = c^2$

Practice A

1. $a^2 + b^2 = c^2$ **2.** $x^2 + z^2 = y^2$
3. $o^2 + m^2 = n^2$ **4.** $2\sqrt{3}$ **5.** $4\sqrt{3}$ **6.** $2\sqrt{5}$
7. $3\sqrt{2}$ **8.** $2\sqrt{15}$ **9.** $5\sqrt{3}$ **10.** 10, yes
11. 12, yes **12.** $4\sqrt{5}$, no **13.** $2\sqrt{29}$, no
14. 12, yes **15.** $\sqrt{89}$, no **16.** 54 cm^2
17. 6 in.^{2} **18.** 60 ft^2 **19.** 33.9 in. **20.** 7.2 ft
21. 9.5 ft

Practice B

1. true **2.** true **3.** false **4.** false **5.** true
6. true **7.** $2\sqrt{3}$, no **8.** 5, yes **9.** $\sqrt{61}$, no
10. 26, yes **11.** 8, no **12.** $3\sqrt{21}$, no
13. 25 cm^2 **14.** 45.3 in.^{2} **15.** 168 cm^2
16. $400 + 100\sqrt{2} \approx 541.4$ mi,
$400 - 100\sqrt{2} \approx 258.6$ mi **17.** 130.3 ft

Practice C

1. $\sqrt{13}$, no **2.** $2\sqrt{5}$, no **3.** $8\sqrt{7}$, no
4. $2\sqrt{11}$, no **5.** 5, no **6.** $6\sqrt{2}$, no **7.** 10
8. 10 **9.** 34 **10.** 45 **11.** 161 cm^2
12. 94.5 in.^{2} **13.** 264.6 ft^2 **14.** 7.3 ft

15. First find the length of each leg:
$AC = |y_2 - y_1|$ and $CB = |x_2 - x_1|$. Now use the Pythagorean Theorem: $AB^2 = AC^2 + BC^2 \Longrightarrow$
$AB^2 = (y_2 - y_1)^2 + (x_2 - x_1)^2 \Longrightarrow$
$AB = \sqrt{(y_2 - y_1)^2 + (x_2 - x_1)^2} \Longrightarrow$
$AB = \sqrt{(x_2 - x_1)^2 + (y_2 - y_1)^2}$ **16.** $\sqrt{5}$

Reteaching with Additional Practice

1. 10.6; no **2.** 20; yes **3.** 13; yes
4. 11.6 units **5.** 3 units **6.** 5.0 units
7. 20.1 square units **8.** 98.8 square units
9. 9.2 square units

Real-Life Application

1. 24 feet per second **2.** 103 feet per second
3. 1.875 seconds **4.** 1.24 seconds **5.** yes

Challenge: Skills and Applications

1. 33 **2.** 14 **3.** $TU = 6, TV = \sqrt{85}$
4. 7056. **5. a.** $a^2 + b^2$ **b.** *Sample answer:* Since $\angle QPX$, $\angle QRW$, $\angle WTU$, and $\angle XVU$ are right angles, they are all congruent. Also, $QP = QR = WT = XV = a$, and $PX = RW = TU = VU = b$. So, the triangles are all congruent by the SAS Congruence Postulate.

c. *Sample answer:* The original figure, with area $a^2 + b^2$, can be sliced and rearranged to form $QWUX$, which is a square with area c^2. So, $a^2 + b^2 = c^2$.

6. a. $\frac{1}{4}(a^2b^2 + b^2c^2 + a^2c^2)$
b. $\frac{1}{4}(a^2 + b^2)(c^2 + k^2)$
c. $\frac{1}{2}ab, \frac{1}{2}k\sqrt{a^2 + b^2}; ab = k\sqrt{a^2 + b^2};$
$a^2b^2 + b^2c^2 + a^2c^2$
$\quad = \left(k\sqrt{a^2 + b^2}\right)^2 + b^2c^2 + a^2c^2$
$\quad = k^2a^2 + k^2b^2 + b^2c^2 + a^2c^2$
$\quad = (a^2 + b^2)(c^2 + k^2)$
So, $\frac{1}{4}(a^2b^2 + b^2c^2 + a^2c^2) = \frac{1}{4}(a^2 + b^2)(c^2 + k^2)$.

Lesson 9.3

Lesson 9.3

Warm-Up Exercises

1. 10   **2.** 5   **3.** 18.9   **4.** 12.6   **5.** 14

Daily Homework Quiz

1. 45   **2.** yes   **3.** 306 m^2

Lesson Opener

Allow 10 minutes.

1. *Sample answer:*

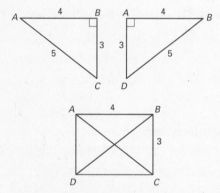

Students could also use color to show the different steps on a single diagram.

2. The 3-4-5 method relies on the converse. Since $3^2 + 4^2 = 5^2$, the triangle formed by lengths 3, 4, and 5 is a right triangle and you have a square corner at the angle formed where the sides of length 3 and length 4 meet.   **3.** *Sample answer:* Use the 3-4-5 method with the units 30 ft, 40 ft, and 50 ft. (So, start with a base line that is 40 ft long.) The result will be a rectangular area that measures 30 ft by 40 ft. Measure 5 ft off each side that measures 30 ft to get sides that are 25 ft long.

Practice A

1. yes   **2.** yes   **3.** no   **4.** yes   **5.** no   **6.** yes
7. no   **8.** yes   **9.** no   **10.** yes   **11.** no
12. no   **13.** right   **14.** obtuse   **15.** acute
16. acute   **17.** right   **18.** acute   **19.** Rectangle; opposite sides are congruent therefore it is a parallelogram; $7^2 + 24^2 = 25^2$ so one pair of opposite angles are right angles; it follows that the parallelogram is a rectangle.   **20.** Square; the diagonals bisect each other so the·quad. is a parallelogram; the diagonals are congruent so the parallelogram

is a rectangle; $4^2 + 4^2 = \left(4\sqrt{2}\right)^2$, so the diagonals intersect at a right angle to form $\perp$ lines; thus the parallelogram is also a rhombus. A quad. that is both a rectangle and a rhombus is a square.

21. Rhombus; the diagonals bisect each other so the quad. is a parallelogram; $6^2 + 8^2 = 10^2$ so the diagonals intersect to form a right angle so the diagonals are $\perp$; thus the parallelogram must be a rhombus.   **22.** Slope $\overline{AC} = -\frac{5}{3}$, slope $\overline{BC} = \frac{3}{5}$; since $\left(\frac{3}{5}\right)\left(-\frac{5}{3}\right) = -1$, $\overline{AC} \perp \overline{BC}$, so $\angle ACB$ is a right angle, therefore, $\triangle ACB$ is a right $\triangle$ by definition.

23. $(AC)^2 + (BC)^2 = 34 + 34 = 68 = (AB)^2$, so by the Converse of the Pythagorean Thm., $\triangle ABC$ is a right triangle.   **24.** Use Method 2 since you need to know the lengths of the sides.

Practice B

1. no   **2.** yes   **3.** no   **4.** yes   **5.** yes   **6.** yes
7. yes, right   **8.** yes, obtuse   **9.** yes, acute
10. yes, right   **11.** yes, obtuse   **12.** yes, right
13. Kite; $5^2 + 12^2 = 13^2$ so diagonals intersect to form a right angle so diagonals are $\perp$; two pair of adjacent sides are congruent and opposite sides not congruent; thus the quad. is a kite.

14. Square; the diagonals bisect each other so the quad. is a parallelogram; the diagonals are congruent so the parallelogram is a rectangle; $7^2 + 7^2 = \left(7\sqrt{2}\right)^2$ so the diagonals intersect at a right angle to form $\perp$ lines; thus the parallelogram is also a rhombus. A quad. that is both a rectangle and a rhombus is a square.   **15.** Rectangle; opposite sides are congruent therefore it is a parallelogram; $5^2 + 12^2 = 13^2$ so one pair of opposite angles are right angles; it follows that the parallelogram is a rectangle.   **16.** If the diagonal's measure is 20.6 ft or 20 ft 7 inches, then the deck is a square by the Converse of the Pythagorean Thm.

17. 39 ft   **18.** $\approx$ 94 rows

Practice C

1. yes   **2.** yes   **3.** no   **4.** no   **5.** no   **6.** yes
7. yes, right   **8.** yes, obtuse   **9.** yes, acute
10. yes, obtuse   **11.** yes, right   **12.** yes, right

13. Kite; $8^2 + 15^2 = 17^2$ so by the Converse of the Pythagorean Thm. the diagonals are $\perp$, also two pairs of consecutive sides are congruent (use congruent triangles), thus the quad. is a kite.

14. Square; the diagonals bisect each other so the quad. is a parallelogram; the diagonals are congruent so the parallelogram is a rectangle; $8^2 + 8^2 = (8\sqrt{2})^2$ so the diagonals intersect at a right angle to form $\perp$ lines; thus the parallelogram is also a rhombus. A quad. that is both a rectangle and a rhombus is a square. **15.** Rectangle; opposite sides are congruent therefore it is a parallelogram; $8^2 + 14^2 = (2\sqrt{65})^2$ so one pair of opposite angles are right angles; it follows that the parallelogram is a rectangle.

16. 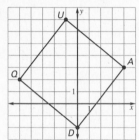 Quad *QUAD* is a square;
perimeter $= 4\sqrt{41}$

17. Since $6^2 + 8^2 = 10^2$, $\triangle ABC$ is a right $\triangle$ by the Converse of the Pythagorean Thm. $\angle ABC$ is a right $\angle$ since it is opposite the longest side. $\angle ABC$ and $\angle 1$ are a linear pair by def. So $\angle 2 = 180° - 90° = 90°$. Therefore, by def. $\angle 2$ is a right $\angle$.

18. Since $3^2 + 6^2 < 8^2$, $\triangle XYZ$ is an obtuse $\triangle$ by Thm. 9.7. $\angle XYZ$ is obtuse since it is opposite the longest side. Since vertical angles are congruent, $\angle XYZ \cong \angle 2$. So by substitution, $\angle 2$ is obtuse.

Reteaching with Additional Practice

1. yes **2.** no **3.** no **4.** yes; right **5.** no
6. yes; obtuse **7.** yes; obtuse

Interdisciplinary Application

1. 18.5 ft; yes **2.** 15 ft; yes **3.** Kitchen 1: sink and refrigerator; Kitchen 2: sink and refrigerator, stove and sink

4. not a right triangle; $4^2 + 5.5^2 \neq 8.7^2$

5. right triangle; $6^2 + 2.5^2 = 6.5^2$

6. It changes the perimeter by 0.2 feet and improves the sink to refrigerator distance by 0.3

feet; not a right triangle.

Challenge: Skills and Applications

1. a. *Sample answer:* $a^2 + b^2 = (n^2 - m^2)^2 + (2mn)^2 = (n^4 - 2n^2m^2 + m^4) + 4m^2n^2 = n^4 + 2m^2n^2 + m^4 = (n^2 + m^2)^2 = c^2$ **b.** 3, 4, 5; 8, 6, 10; 15, 8, 17; 24, 10, 26; 5, 12, 13; 12, 16, 20; 21, 20, 29; 7, 24, 25; 16, 30, 34; 9, 40, 41

c. $m = \sqrt{\dfrac{c-a}{2}}$; $n = \sqrt{\dfrac{c+a}{2}}$

d. *Sample answer:* Let c be the largest number; choose a so that $\dfrac{c-a}{2}$ and $\dfrac{c+a}{2}$ are perfect squares. **e.** $m = 5, n = 9$

f. $m = 3, n = 8$ **2. a.** *Sample answer:* $\overline{RS}$ can be extended because two points determine a line; lines perpendicular to $\overleftrightarrow{RS}$ that pass through Q and through P can be constructed by the Perpendicular Postulate. **b.** $2c^2 + 2d^2$ **c.** *Sample answer:* The sum of the squares of the diagonal lengths is equal to the sum of the squares of the side lengths.

d. *Sample answer:* Since *ABCD* is a kite, the diagonals are perpendicular, so $x^2 + z^2 = u^2$ and $y^2 + z^2 = v^2$. If the relationship in part (c) holds, we have:

$$(x + y)^2 + (2z)^2 = 2u^2 + 2v^2$$
$$x^2 + 2xy + y^2 + 4z^2 = 2(x^2 + z^2) + 2(y^2 + z^2)$$
$$0 = x^2 - 2xy + y^2$$
$$0 = (x - y)^2$$

Thus, $x = y$ and the diagonals bisect each other, contradicting the given information that *ABCD* is a kite. So, if *ABCD* is a kite, the relationship in part (c) does not hold. **3.** 20 **4.** 36, $\sqrt{71}$

5. $8, \frac{1}{3}(10 + \sqrt{46})$

6. $7, 9, 8 + \sqrt{161}, \frac{1}{13}(-4 + 5\sqrt{127})$

7. $\sqrt{31} < x < 16$ **8.** $2 < x < 3$ or $x > 12$

Quiz 1

1. $\triangle RTQ \sim \triangle RSQ \sim \triangle QST$ **2.** $\overline{SQ}$ **3.** 6
4. $2\sqrt{3} \approx 3.46$ **5.** $4\sqrt{2}$ **6.** 12 **7.** $4\sqrt{35}$

8. no; $\dfrac{\sqrt{90^2 + 90^2}}{90} \neq \dfrac{90}{60\frac{1}{2}}$

Lesson 9.4

Lesson 9.4

Warm-Up Exercises

1. $6\sqrt{2}$ **2.** $4\sqrt{2}$ **3.** 20 **4.** $8\sqrt{3}$ **5.** 18

Daily Homework Quiz

1. yes; right **2.** yes; obtuse **3.** no **4.** acute
5. right

Lesson Opener

Allow 20 minutes.

1. The area of one square is twice the area of the next smaller square. **2.** 8 different sizes; 4 of each size; 45°, 45°, 90°; the lengths are equal.

3.

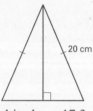

4. Answers will vary.

Practice A

1. $x = 60°, y = 30°$ **2.** $x = 45°, y = 45°$
3. $x = 60°, y = 30°$ **4.** $x = 8, y = 4\sqrt{3}$
5. $x = 4\sqrt{2}, y = 4\sqrt{2}$ **6.** $x = 10, y = 10\sqrt{2}$
7. $x = 3, y = 3\sqrt{3}$ **8.** $x = 24, y = 12\sqrt{3}$
9. $x = 6\sqrt{2}, y = 6\sqrt{2}$

10.   **11.**

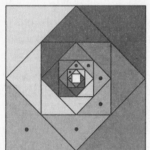

altitude $\approx$ 17.3 cm   diagonal $\approx$ 7.1 cm

12.   **13.** $\approx$ 127.3 ft
                          14. $\approx$ 127.3 ft
         15. no

side $\approx$ 7.1 inches

Practice B

1. $x = 8\sqrt{3}, y = 16$ **2.** $x = 5\sqrt{2}, y = 5\sqrt{2}$
3. $x = 7, y = 7\sqrt{2}$ **4.** $x = \dfrac{8\sqrt{3}}{3}, y = \dfrac{16\sqrt{3}}{3}$
5. $x = \dfrac{10\sqrt{3}}{3}, y = \dfrac{20\sqrt{3}}{3}$
6. $x = 13, y = 13\sqrt{2}$ **7.** $x = 20\sqrt{3}, y = 40$
8. $x = 9\sqrt{2}, y = 9\sqrt{2}$
9. $x = 5\sqrt{3}, y = 10\sqrt{3}$

10. diagonal $\approx$ 7.1 cm   **11.** side $\approx$ 20.8 inches

12. side $\approx$ 11.3 cm   **13.** side $= \frac{5}{2}$

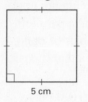

14. $3000\sqrt{3}$ ft $\approx$ 5196.2 ft
15. $3000\sqrt{3}$ ft $\approx$ 5196.2 ft **16.** less

Practice C

1. $x = 12, y = 8\sqrt{3}$ **2.** $x = 6, y = 6$
3. $x = 3\sqrt{3}, y = 9$ **4.** $x = 8, y = 8\sqrt{3}$
5. $x = 24, y = 12\sqrt{3}$ **6.** $x = 15\sqrt{2}, y = 15$
7. $x = 10\sqrt{3}, y = 15$ **8.** $x = 8, y = 8\sqrt{2}$
9. $x = 4, y = \dfrac{8\sqrt{3}}{3}$

10. perimeter $\approx$ 41.6 cm   **11.** area $= 32$ in.^{2}

12.   **13.**
diagonal $\approx$ 24.6 cm   altitude $\approx$ 10.4 inches

          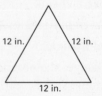

Lesson 9.4 *continued*

14. 2.25 **15.** 1

Reteaching with Additional Practice

1. $42\sqrt{2} \approx 59.4$ **2.** $9\sqrt{2} \approx 12.7$

3. $x = \dfrac{13\sqrt{2}}{2} \approx 9.2, y = \dfrac{13\sqrt{2}}{2} \approx 9.2$

4. $14\sqrt{3} \approx 24.2$ **5.** $7\sqrt{3} \approx 12.1$

6. $x = 5\sqrt{3} \approx 8.7, y = 10\sqrt{3} \approx 17.4$

Real-Life Application

1.

2. 45° **3.** 6.9 feet

4. 20.8 feet **5.** 60° pitch **6.** 30° pitch

Math & History Application

1. $\frac{1}{2}(a + b)(a + b)$ **2.** $\frac{1}{2}ab + \frac{1}{2}ab + \frac{1}{2}c^2$

3. $\frac{1}{2}(a + b)(a + b) = \frac{1}{2}ab + \frac{1}{2}ab + \frac{1}{2}c^2$
$(a + b)(a + b) = ab + ab + c^2$
$a^2 + 2ab + b^2 = 2ab + c^2$
$a^2 + b^2 = c^2$

Challenge: Skills and Applications

1. $p = \sqrt{3}, q = 2\sqrt{3}, r = 2, s = 4, t = 4,$
$u = 4\sqrt{2}, v = 2\sqrt{2}, w = 2\sqrt{6}$

2. $\dfrac{9}{2}\sqrt{3} \approx 7.79$ in. **3.** $\dfrac{b - a}{2}$

4. $\frac{1}{2}(b - a)(3 - \sqrt{3})$

5. a. $XY = 2\sqrt{3}, VY = 2\sqrt{3}, VZ = 2\sqrt{3} - 2$
b. 30°-60°-90° right triangle; $VW = 3 - \sqrt{3}$,
$WZ = \sqrt{3} - 1, VZ = 2\sqrt{3} - 2$
c. $VW = 3 - \sqrt{3}, WX = 3 + \sqrt{3}$,
$VX = 2\sqrt{6}$

6. a. $AB = 1 - x, AD = 2$
b. $\dfrac{2}{1 - x} = \dfrac{\sqrt{3}}{x}; x = 2\sqrt{3} - 3$
c. $BC = 2\sqrt{3} - 3, CD = \sqrt{3}$,
$BD = 2\sqrt{6 - 3\sqrt{3}}$

7. *Sample answer:*

$\dfrac{BC}{WV} = \dfrac{CD}{WX} = \dfrac{BD}{VX} = \dfrac{\sqrt{3} - 1}{2} \approx 0.366$; by the SSS Similarity Theorem, the triangles are similar.

Lesson 9.5

Warm-Up Exercises

1. 15.75 **2.** 48 **3.** 24 **4.** 13.9 **5.** 36

Daily Homework Quiz

1. $x = 7, y = 14$ **2–3.** Check students' sketches. **2.** $12\sqrt{2}$ **3.** 43.3 square inches

Lesson Opener

Allow 10 minutes.

1. No; Yes **2.** Answers will vary. **3.** They are approximately equal; the triangles are similar, so ratios of corresponding lengths should be equal; no. **4.** Yes. *Sample answer:* Groups chose different measures for $\angle A$, which resulted in different ratios.

Practice A

1. $\dfrac{a}{c}$ **2.** $\dfrac{b}{c}$ **3.** $\dfrac{b}{a}$ **4.** $\dfrac{j}{l}$ **5.** $\dfrac{j}{l}$ **6.** $\dfrac{k}{j}$

7. $\sin D = 0.8, \cos D = 0.6, \tan D \approx 1.3333,$
$\sin F = 0.6, \cos F = 0.8, \tan F = 0.75$

8. $\sin D = \cos D = \sin F = \cos F \approx 0.7071,$
$\tan D = \tan F = 1$

9. $\sin D \approx 0.9231, \cos D \approx 0.3846,$
$\tan D = 2.4, \sin F \approx 0.3846, \cos F \approx 0.9231,$
$\tan F \approx 0.4167$

10. 0.5 **11.** 0.9511 **12.** 3.0777 **13.** 0.7431

14. 0.9004 **15.** 0.4226 **16.** 0.2493

17. 0.9925 **18.** 15; 11.8 **19.** 37°; 11.2

20. x; 18.0 **21.** $x \approx 5.9, y \approx 8.1$

22. $x \approx 3.5, y \approx 7.2$ **23.** $x \approx 8.9, y \approx 12.0$

Practice B

1. $\sin M \approx 0.5812, \cos M \approx 0.8137,$
$\tan M \approx 0.7143, \sin T \approx 0.8137, \cos T \approx 0.5812,$
$\tan T = 1.4$

2. $\sin A = \cos A = \sin N = \cos N \approx 0.7071,$
$\tan A = \tan N = 1$

Lesson 9.5 *continued*

3. $\sin Q = 0.8$, $\cos Q = 0.6$, $\tan Q \approx 1.3333$, $\sin R = 0.6$, $\cos R = 0.8$, $\tan R = 0.75$

4. $\sin O \approx 0.5039$, $\cos O \approx 0.8638$, $\tan O \approx 0.5833$, $\sin T \approx 0.8638$, $\cos T \approx 0.5039$, $\tan T \approx 1.7143$

5. $\sin B = \cos B = \sin X = \cos X \approx 0.7071$, $\tan B = \tan X = 1$

6. $\sin B \approx 0.4679$, $\cos B \approx 0.8838$, $\tan B \approx 0.5294$, $\sin A \approx 0.8838$, $\cos A \approx 0.4679$, $\tan A \approx 1.8889$

7. 0.1736   **8.** 0.7880   **9.** 0.9657   **10.** 0.9613

11. 2.1445   **12.** 0.4540   **13.** 0.8387

14. 0.8387   **15.** $x \approx 9.2$, $y \approx 9.7$

16. $x \approx 10.4$, $y \approx 6.7$ **17.** $x \approx 4.7$, $y \approx 11.0$

18. $x \approx 1.7$, $y \approx 4.7$ **19.** $x \approx 5.4$, $y \approx 10.7$

20. $x \approx 23.9$, $y \approx 29.9$ **21.** about 184.3 ft

Practice C

1. $\sin T \approx 0.5547$, $\cos T \approx 0.8321$, $\tan T \approx 0.6667$, $\sin A \approx 0.8321$, $\cos A \approx 0.5547$, $\tan A = 1.5$

2. $\sin J = \cos J = \sin E = \cos E \approx 0.7071$, $\tan J = \tan E = 1$

3. $\sin A = 0.28$, $\cos A = 0.96$, $\tan A \approx 0.2917$, $\sin B = 0.96$, $\cos B = 0.28$, $\tan B \approx 3.4286$

4. 0.7547   **5.** 0.1219   **6.** 0.0699   **7.** 0.9455

8. 3.7321   **9.** 0.9659   **10.** 0.5299

11. 0.4384   **12.** $x \approx 7.8$, $y \approx 16.2$

13. $x \approx 10.9$, $y \approx 17.8$ **14.** $x \approx 5.2$, $y \approx 15.9$

15. about 67.7 ft   **16.** about 80.7 ft

17. Longer; as the sun sets the angle decreases and the tangent of the angle also decreases. The height of the lighthouse is constant so the shadow has to lengthen for the ratio to get smaller.

18. about 25.0 ft   **19.** Shortens by about 1.2 feet; if the angle increases, so does the sine of the angle. The vertical distance between floors is constant so the length of the escalator must shorten for the ratio to increase.

Reteaching with Additional Practice

1. $\sin A = 0.8$, $\cos A = 0.6$, $\tan A = 1.3333$

2. $\sin A = 0.7667$, $\cos A = 0.6333$, $\tan A = 1.2105$

3. $\sin A = 0.8333$, $\cos A = 0.5556$, $\tan A = 1.5$

4. 35.2 feet   **5.** 70.7 meters

Cooperative Learning Activity

1. Answers may vary.   **2.** Tangent

3. $\tan 90°$ is undefined.

Interdisciplinary Application

1. 9.78 meters   **2.** about 79.92°   **3.** 10.08°

4. 5.2°   **5.** about 1.67°   **6.** 55.839 meters

Challenge: Skills and Applications

1. 21.0   **2. a.** $x \tan 62°$, $(x + 25) \tan 40°$

b. 37.9 ft   **3.** $JK \approx 40.1$, $KL \approx 27.2$

4. $\tan x° = \dfrac{a}{b}$, $\tan(90 - x)° = \dfrac{b}{a}$; they are reciprocals.

5. $(\sin x°)^2 + (\cos x°)^2 = \left(\dfrac{a}{c}\right)^2 + \left(\dfrac{b}{c}\right)^2 = \dfrac{a^2 + b^2}{c^2} = \dfrac{c^2}{c^2} = 1$

6. 0.8   **7. a.** *Sample answer:* Since $PT = QT$, we have $\angle PQT \cong \angle P$, so $m\angle PQT = 36°$. So by the Exterior Angle Theorem, $m\angle QTR = m\angle P + m\angle PQT = 36° + 36° = 72°$. Applying the Base Angles Theorem again, since $QT = QR$, we have $m\angle R = m\angle QTR = 72°$, so $m\angle RQT = 180° - 72° - 72° = 36°$. So $\angle RQT \cong \angle RPQ$, and $\angle R \cong \angle R$ (Reflexive Property of Congruence); therefore, $\triangle PRQ \sim \triangle QRT$ (AA Similarity Postulate).

b. $\dfrac{1}{2x} = \dfrac{1 + 2x}{1}$; $x = \dfrac{-1 \pm \sqrt{5}}{4}$

c. $\angle SQT$ (or $\angle RQS$); $\dfrac{-1 + \sqrt{5}}{4}$

Quiz 2

1. 5.2 m   **2.** 11.3 in.   **3.** 43.3 m²

4. $x \approx 23.4$, $y \approx 21.9$ **5.** $x \approx 9.2$, $y \approx 7.7$

6. $x \approx 7.0$, $y \approx 12.2$ **7.** ≈ 84 ft

Lesson 9.6

Lesson 9.6

Warm-Up Exercises

1. ≈ 0.3846 2. ≈ 0.9231 3. ≈ 0.4167
4. ≈ 0.9231 5. ≈ 0.3846 6. $= 2.4$

Daily Homework Quiz

1. $\sin B \approx 0.9459$, $\cos B \approx 0.3243$,
$\tan B \approx 2.9167$, $\sin C \approx 0.3243$,
$\cos C \approx 0.9459$, $\tan C \approx 0.3429$
2. $x \approx 5$ m, $y \approx 9.4$ m 3. about 20 m^2

Lesson Opener

Allow 10 minutes.

1.

2. $\tan A = \dfrac{3}{12}$; $m\angle A \approx 14°$ 3. Steeper; this

roof rises 2 more inches for every 12 inches
across than the first roof does. Students can also
explain by using the Table on p. 845 to find the
measure of the angle of this roof. Tan $A = \frac{5}{12}$
(≈ 0.4167), when $m\angle A$ is between 22° and 23°.

Practice A

1. A 2. B 3. C 4. B 5. A 6. D
7. 8.9 8. 48.2° 9. 41.8° 10. 24.8°
11. 68.2° 12. 11.5° 13. 1.1° 14. 47.2°
15. 33.0° 16. 29.3° 17. 56.3°
18. $BC \approx 8.9$, $m\angle A \approx 48.2°$, $m\angle B \approx 41.8°$
19. $RT \approx 13.6$, $m\angle T \approx 36.0°$, $m\angle R \approx 54.0°$
20. $ML \approx 6.6$, $MN \approx 12.4$, $m\angle L = 62°$
21. $m\angle T = 42°$, $PT \approx 14.4$, $TQ \approx 19.4$
22. $m\angle D = 71°$, $EF \approx 61.0$, $FD \approx 64.5$
23. $m\angle J = 49°$, $IK \approx 19.6$, $IJ \approx 25.9$
24. about 75.5°

Practice B

1. 16.6 2. 65.0° 3. 25.0° 4. 13.9°
5. 60.0° 6. 51.7° 7. 4.6° 8. 19.9°
9. 41.0° 10. 22.3° 11. 69.4°
12. $m\angle P = 53°$, $PQ \approx 13.2$, $QR \approx 17.6$

13. $m\angle P \approx 58.6°$, $m\angle N \approx 31.4°$, $PN \approx 21.1$
14. $TU \approx 21.9$, $m\angle S \approx 72.3°$, $m\angle U \approx 17.7°$
15. $m\angle V = 39°$, $DM \approx 11.3$, $DV \approx 18.0$
16. $m\angle T = 66°$, $TR \approx 14.7$, $AT \approx 36.1$
17. $UM \approx 20.6$, $m\angle U \approx 42.7°$, $m\angle E \approx 47.3°$
18. about 7.5 ft 19. about 84.02 ft
20. about 499.30 ft

Practice C

1. 58.2° 2. 64.9° 3. 29.5° 4. 3.4°
5. 81.4° 6. 46.4° 7. 24.8° 8. 40.0°
9. $AB \approx 17.9$, $m\angle A \approx 63.4°$, $m\angle B \approx 26.6°$
10. $m\angle J = 37°$, $IJ \approx 18.4$, $HJ \approx 13.8$
11. $m\angle M = 63°$, $MO \approx 6.6$, $MP \approx 14.6$
12. $ST \approx 17.2$, $m\angle T \approx 57.5°$, $m\angle R \approx 32.5°$
13. $m\angle Z = 62.5°$, $XY \approx 15.8$, $XZ \approx 17.8$
14. $RT \approx 11.9$, $m\angle R \approx 51.5°$, $m\angle T \approx 38.5°$
15. $m\angle D = 59°$, $DE \approx 10.5$, $DF \approx 20.4$
16. $MN \approx 25.1$, $m\angle N \approx 70.9°$, $m\angle M \approx 19.1°$
17. $m\angle T = 34°$, $NC \approx 10.4$, $CT \approx 18.6$
18. about 229.4 m 19. about 63.8 ft

Reteaching with Additional Practice

1. $x = 14$, $y \approx 20.7$, $z = 25$, $Y = 55.9°$,
$X = 34.1°$, $Z = 90°$ 2. $l = 14$, $m = 12$,
$n \approx 18.4$, $L = 49.4°$, $M = 40.6°$, $N = 90°$
3. $p = 9.4$, $q = 5.3$, $r = 10.8$, $Q = 29.5°$,
$P = 60.5°$, $R = 90°$ 4. $a = 19$, $b = 30.9$,
$c = 24.3$, $A = 38°$, $B = 90°$, $C = 52°$
5. $l = 5.2$, $m = 9$, $n = 7.3$, $L = 35°$, $M = 90°$,
$N = 55°$ 6. $p = 55.9$, $q = 41.5$, $r = 37.4$,
$P = 90°$, $Q = 48°$, $R = 42°$

Real-Life Application

1. 12.0 units 2. 32.9 units 3. 17.1 units
4. Pythagorean Theorem 5. 209,000 light years

Challenge: Skills and Applications

1. a. $b \sin A$ b. $a \sin B$

c. $b \sin A = a \sin B$, so $\dfrac{\sin A}{a} = \dfrac{\sin B}{b}$.

2. $b \approx 31.1$, $c \approx 34.0$, $m\angle C = 75°$
3. $a \approx 20.4$, $c \approx 18.2$, $m\angle C = 49°$
4. $b \approx 24.1$, $m\angle A = m\angle C = 48°$

Lesson 9.6 *continued*

5. a. $p^2 + q^2 = b^2$; $p = b \cos A$

b. $a^2 = c^2 - 2cp + p^2 + q^2$

c. $a^2 = c^2 - 2bc \cos A + b^2$

6. $m\angle A \approx 44.9°$, $m\angle B \approx 79.1°$, $m\angle C \approx 56.0°$

7. $b \approx 90.8$, $m\angle A \approx 28.1°$, $m\angle C \approx 79.9°$

8. $m\angle A \approx 56.6°$, $m\angle B = 90°$, $m\angle C \approx 33.4°$

Lesson 9.7

Warm-Up Exercises

1. 5 **2.** 5 **3.** $\sqrt{26}$ **4.** $4\sqrt{2}$ **5.** $\sqrt{29}$

Daily Homework Quiz

1. 14.5° **2.** 67.7° **3.** side lengths: 4, 5, 6.4; angle measures: 90°, 38.7°, 51.3° **4.** side lengths: 7, 4.9, 8.5; angle measures: 90°, 35°, 55°

Lesson Opener

Allow 5 minutes.

1. Car A: 15 mi/h north, 30 mi/h north; Car B: 5 mi/h northeast, 25 mi/h east. **2.** Car A is traveling three times as fast as Car B before the intersections; Length of vector for Car A is three times length of vector for Car B. **3.** Check drawings. Vector for Car A should point up and should be twice the length of vector for Car A shown in Question 2. Vector for Car B should point to the right and should be five times as long as the vector for Car B shown in Question 2.

Technology Activity

1. 10.77 knots **2.** 22° with the positive *x*-axis

3. 11.66 knots at an angle of 31° with the positive *x*-axis **4.** Answers should include a reference to the tangent of the angle made between the sum vector and the positive *x*-axis and the value of the slope.

Practice A

1. C **2.** A **3.** B **4.** $\langle 2, 2 \rangle$; 2.8

5. $\langle 4, 3 \rangle$; 5 **6.** $\langle -4, 4 \rangle$; 5.7

7.

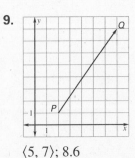

$\langle 3, 4 \rangle$; 5

8.

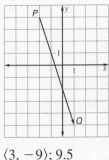

$\langle 5, 8 \rangle$; 9.4

9.
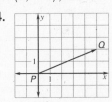

$\langle 5, 7 \rangle$; 8.6

10.

$\langle 3, -9 \rangle$; 9.5

11. about 73 mph; about 16° north of east

12. about 78 mph; about 50° south of east

Practice B

1. $\langle 3, 5 \rangle$; 5.8 **2.** $\langle -5, -5 \rangle$; 7.1

3. $\langle 7, -4 \rangle$; 8.1

4.
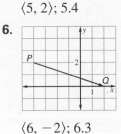

$\langle 5, 2 \rangle$; 5.4

5.
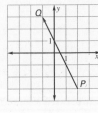

$\langle 4, -4 \rangle$; 5.7

6.

$\langle 6, -2 \rangle$; 6.3

7.

$\langle -3, 6 \rangle$; 6.7

8. about 89 mph; about 27° north of east

9. about 73 mph; about 16° south of west

10. $\overrightarrow{IJ}$, $\overrightarrow{AB}$, $\overrightarrow{CD}$ **11.** $\overrightarrow{AB}$ and $\overrightarrow{IJ}$

12. $\overrightarrow{AB}$ and $\overrightarrow{IJ}$ **13.** $\overrightarrow{CD}$ and $\overrightarrow{GH}$

Practice C

1. $\langle 7, 4 \rangle$; 8.1 **2.** $\langle 6, -7 \rangle$; 9.2

3. $\langle -7, -6 \rangle$; 9.2

Lesson 9.7 *continued*

4. about 94 mph; about 32° south of west

5. about 157 mph; about 17° north of east

6. $\vec{u} = \langle 3, 2 \rangle$, $\vec{v} = \langle 2, 4 \rangle$,
$\vec{u} + \vec{v} = \langle 5, 6 \rangle$

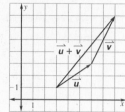

7. $\vec{u} = \langle 6, -5 \rangle$,
$\vec{v} = \langle -3, -4 \rangle$,
$\vec{u} + \vec{v} = \langle 3, -9 \rangle$

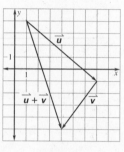

8. $\langle 5, -5 \rangle$ **9.** $\langle 7, -3 \rangle$ **10.** $\langle 6, -4 \rangle$

Reteaching with Additional Practice

1. $\langle -2, 7 \rangle$; ≈ 7.3 **2.** $\langle 2, -6 \rangle$; ≈ 6.3

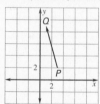 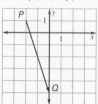

3. $\langle -4, 2 \rangle$; ≈ 4.5 **4.** $\langle 4, 13 \rangle$; ≈ 13.6

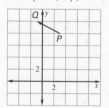 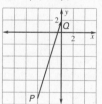

5. 9.2, 49.4° north of east

6. 7.1, 45° south of east

7. 7.1, 45° south of west

8. $\langle -3, 13 \rangle$ **9.** $\langle 0, 3 \rangle$ **10.** $\langle 0, 0 \rangle$

Interdisciplinary Application

1. $x = 62.23t$, $y = 62.23t - 16t^2$

2. $x = 82.69t$, $y = 30.10t - 16t^2$

3. Team 2; Team 2 **4.** Team 1: 3.9 seconds;
Team 2: 1.9 seconds **5.** Team 1; 85.6 feet

Challenge: Skills and Applications

1. $(\vec{u} + \vec{v}) + \vec{w}$
$$= \langle a_1 + a_2, b_1 + b_2 \rangle + \langle a_3, b_3 \rangle$$
$$= \langle (a_1 + a_2) + a_3, (b_1 + b_2) + b_3 \rangle$$
$$= \langle a_1 + (a_2 + a_3), b_1 + (b_2 + b_3) \rangle$$
$$= \langle a_1, b_1 \rangle + \langle a_2 + a_3, b_2 + b_3 \rangle$$
$$= \vec{u} + (\vec{v} + \vec{w})$$

2. $\vec{u} \cdot \vec{v} = a_1 a_2 + b_1 b_2$
$$= a_2 a_1 + b_2 b_1$$
$$= \vec{v} \cdot \vec{u}$$

3. $k(\vec{u} + \vec{v}) = k\langle a_1 + a_2, b_1 + b_2 \rangle$
$$= \langle k(a_1 + a_2), k(b_1 + b_2) \rangle$$
$$= \langle ka_1 + ka_2, kb_1 + kb_2 \rangle$$
$$= \langle ka_1, kb_1 \rangle + \langle ka_2, kb_2 \rangle$$
$$= k\vec{u} + k\vec{v}$$

4. $\vec{u} \cdot (\vec{v} + \vec{w})$
$$= \langle a_1, b_1 \rangle \cdot \langle a_2 + a_3, b_2 + b_3 \rangle$$
$$= a_1(a_2 + a_3) + b_1(b_2 + b_3)$$
$$= (a_1 a_2 + a_1 a_3) + (b_1 b_2 + b_1 b_3)$$
$$= (a_1 a_2 + b_1 b_2) + (a_1 a_3 + b_1 b_3)$$
$$= \vec{u} \cdot \vec{v} + \vec{u} \cdot \vec{w}$$

5. $(k\vec{u}) \cdot \vec{v} = \langle ka_1, kb_1 \rangle \cdot \langle a_2, b_2 \rangle$
$$= \langle (ka_1)a_2, (kb_1)b_2 \rangle$$
$$= \langle k(a_1 a_2), k(b_1 b_2) \rangle$$
$$= k(a_1 a_2 + b_1 b_2)$$
$$= k(\vec{u} \cdot \vec{v})$$

6. $\langle 2, 5 \rangle$ **7.** $\langle 35, 21 \rangle$ **8.** -24 **9.** $\langle 11, -4 \rangle$

10. $\langle -23, 32 \rangle$ **11.** 9 **12.** -57 **13.** 18

Review and Assessment

Review Games and Activities

Triangle 1   Triangle 2

1. 6 **2.** 7 **3.** 2 **4.** 5 **5.** 6 **6.** 3 **7.** 1

8. 5 **9.** 1 **10.** 4 **11.** 8 **12.** 5

Review and Assessment *continued*

Test A

1. 8; 4 2. 16; 8 3. 12; 6 4. $\sqrt{58}$; no

5. 12; yes 6. $4\sqrt{21}$; no 7. right 8. obtuse

9. acute 10. obtuse 11. no 12. right

13. $x = 6, y = 6\sqrt{2}$ 14. $a = \dfrac{5}{2}, b = \dfrac{5\sqrt{3}}{2}$

15. $m = 6\sqrt{2}, n = 6\sqrt{2}$

16. 0.2195; 0.9756; 0.225 17. 0.6; 0.8; 0.75

18. 0.3846; 0.9231; 0.4167 19. 19.7 20. 11.9

21. $\langle 2, 4 \rangle$; 4.5 22. $\langle 4, 2 \rangle$; 4.5

23. $\langle -2, -3 \rangle$; 3.6 24. $\langle -5, 4 \rangle$; 6.4

25. $\langle 0, 0 \rangle$ 26. $\langle 7, 7 \rangle$ 27. $\langle 8, 9 \rangle$ 28. $\langle 2, -5 \rangle$

29. $\langle 5, 12 \rangle$ 30. $\langle -1, -2 \rangle$

Test B

1. 12; 18 2. 6; 13 3. $4x$; 8 4. 5; yes

5. 9; yes 6. 3.4; no 7. acute 8. obtuse

9. acute 10. right 11. no 12. right

13. $x = 9, y = 9\sqrt{2}$ 14. $x = 16, y = 8\sqrt{3}$

15. $x = 6\sqrt{2}, y = 6\sqrt{2}$

16. 0.3846; 0.9231; 0.4167 17. 0.6; 0.8; 0.75

18. 0.2774; 0.9608; 0.2887 19. 14 20. 8.6

21. $\langle 5, 3 \rangle$; 5.8 22. $\langle 5, -5 \rangle$; 7.1

23. $\langle -4, 0 \rangle$; 4 24. $\langle 8, 3 \rangle$; 8.5

25. $\langle 2, 11 \rangle$ 26. $\langle -4, -1 \rangle$ 27. $\langle -6, -1 \rangle$

28. $\langle -2, -2 \rangle$ 29. $\langle 0, 12 \rangle$ 30. $\langle 4, 11 \rangle$

Test C

1. 7; $\dfrac{49}{3}$ 2. 8; $\dfrac{64}{3}$ 3. $2x$; $5\sqrt{2}$ 4. 2; no

5. 14; no 6. 16.5; no 7. right 8. acute

9. no 10. right 11. obtuse 12. right

13. $x = 5, y = 5\sqrt{2}$ 14. $x = 5\sqrt{3}, y = 10$

15. $x = 5\sqrt{3}, y = 5$ 16. 0.6; 0.8; 0.75

17. 0.6; 0.8; 0.75 18. 0.4854; 0.8738; 0.5556

19. 7.2 20. 8.8 21. $\langle -3, -8 \rangle$; 8.5

22. $\langle -2, -2 \rangle$; 2.8 23. $\langle 5, 14 \rangle$; 14.9

24. $\langle -3, -1 \rangle$; 3.2 25. $\langle -0.75, -12 \rangle$

26. $\langle -3.5, 5.75 \rangle$ 27. $\langle -7, -4 \rangle$

28. $\langle -1.75, -2 \rangle$ 29. $\langle -2.5, -4.25 \rangle$

30. $\langle 2.75, -2.25 \rangle$

SAT/ACT Test

1. A 2. C 3. B 4. E 5. D 6. B 7. A

8. C 9. E 10. B

Alternative Assessment

2. a. and b.

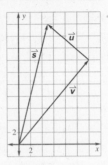

Project: Proving a Conjecture

1–2.

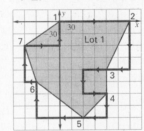 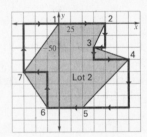

3. Perimeter of Lot 1: 812.05 ft;
Perimeter of Lot 2: 666.52 ft
Area of Lot 1: 40,050 ft²;
Area of Lot 2: 24,687.5 ft²

4. Lot 1: $5481.34; Lot 2: $4499.01

5. Lot 1: $628.86; Lot 2: $413.47

6. *Sample answer:* The advantage for Lot 2 is that the cost of taxes and of fencing are both considerably less. The advantage for Lot 1 is that, with a lower tax rate and a proportionally smaller perimeter to fence (about 1.6 times the land with only about 1.2 times the perimeter), you get "more for your money." 7. Check answers. Students should realize that for only an additional $700 in purchase price, parents would be buying a lot that is more than one and a half times as large. However, the taxes on Lot 1 will be one and a half times as much as Lot 2 annually, so they may only choose Lot 1 if the extra land is important to them.

Cumulative Review

1. 12.2 **2.** 11.3 **3. a.** If the $m\angle 1 \neq 110°$, then $\angle 1$ is not an obtuse angle. **b.** If $\angle 1$ is an obtuse angle, then $m\angle 1 = 110°$.

4. a. If the sun is not shining, then it is raining. **b.** If it is not raining, then the sun is shining.

5. Alternate Exterior Angles Converse Theorem

6. Corresponding Angles Converse Postulate

7.

Statements	Reasons
1. $\overline{MN} \cong \overline{PN}$	**1.** Given
2. $\angle M \cong \angle P$	**2.** Base Angles Theorem
3. O is a midpoint of $\overline{MP}$.	**3.** Given
4. $\overline{MO} \cong \overline{PO}$	**4.** Definition of midpoint
5. $\triangle MON \cong \triangle PON$	**5.** SAS Congruence Post.

8.

Statements	Reasons
1. $\overline{GH} \parallel \overline{KJ}$	**1.** Given
2. $\angle H \cong \angle J$	**2.** Alt. Int. Angles Thm.
3. $\overline{LG} \cong \overline{LK}$	**3.** Given
4. $\triangle GHL \cong \triangle KJL$	**4.** AAS Congruence Thm.
5. $\angle H \cong \angle K$	**5.** Corresp. parts of $\cong$ triangles are $\cong$.

9. 8 **10.** 32 **11.** rhombus

12. isosceles trapezoid **13.** $A(1, 5)$, $B(3, 10)$

14. $A(6, 3)$, $B(8, 8)$ **15.** $x = 9, y = 10$

16. $x = 70°, y = 8$ **17.** 4 **18.** 7.2

19. obtuse **20.** right **21.** $x = 3\sqrt{3}, y = 6$

22. $x = 18.0, y = 19.7$

23. $c = 13, m\angle A = 67.4°, m\angle B = 22.6°$

24. $a = 8, m\angle A = 53.1°, m\angle B = 36.9°$

25. $\langle 4, 4 \rangle$, 5.7 **26.** $\langle 4, -3 \rangle$, 5.0